Introduction to Computational Optimization Models for Production Planning in a Supply Chain

Stefan Voß
David L. Woodruff

Introduction to Computational Optimization Models for Production Planning in a Supply Chain

Second Edition
with 36 Figures
and 24 Tables

Springer

Professor Dr. Stefan Voß
Universität Hamburg
Institut für Wirtschaftsinformatik
Von-Melle-Park 5
20146 Hamburg
Germany
E-mail: stefan.voss@uni-hamburg.de

Professor David L. Woodruff, Ph. D.
Graduate School of Management
UC Davis
Davis CA 95616
USA
E-mail: dlwoodruff@ucdavis.edu

Cataloging-in-Publication Data
Library of Congress Control Number: 2005937007

ISBN-10 3-540-29878-9 2nd ed. Springer Berlin Heidelberg New York
ISBN-13 978-3-540-29878-6 2nd ed. Springer Berlin Heidelberg New York
ISBN 3-540-00023-2 1st ed. Springer Berlin Heidelberg New York

Springer is a part of Springer Science+Business Media
springeronline.com

Printed in Germany

Cover design: Erich Kirchner
Production: Helmut Petri
Printing: Strauss Offsetdruck

SPIN 11578666 Printed on acid-free paper – 42/3153 – 5 4 3 2 1 0

Preface of the Second Edition

After using our book for courses in a variety of locations, we have been gratified by the positive responses to the first edition. Moreover, the book had been used by quite a few practitioners in the way as we had hoped for when we wrote the first edition. So our preparations for the second edition were guided by a desire to add new material and to take advantage of our experience with the first edition, but at the same time to preserve the things that had been proven to be most useful.

The result is this second edition. The first seven chapters are very similar to the first edition, except for corrections and the addition of clarification in places that our experience with the first edition demonstrated a need. For the later chapters, on the other hand, we took the liberty of extending our research presentation a bit and, most importantly, updating the references section.

Stefan Voß
David L. Woodruff

October 2005

The Preface of the First Edition

> *"For almost any program, there exists a champion who can make it work – at least for a while."* Hopp and Spearman (2000)

Managers and information technology professionals need to have an understanding of computational optimization models for production planning in a supply chain. This book provides an accessible introduction to the subject. We develop the terminology and concepts needed to understand the important issues.

We are not trying to be all things to all people. In particular, we are not trying to describe algorithms used by commercial software firms, but rather provide models that could be used by do-it-yourselfers and also can be used to provide understanding of the background issues so that one can do a better job of working with the (proprietary) algorithms of the software vendors.

In this book we strive to provide models that capture many of the details faced by firms operating in a modern supply chain, but we stop short of proposing models for economic analysis of the entire multi-player chain. In other words, we produce models that are useful for planning within a supply chain rather than models for planning the supply chain. The usefulness of the models is enhanced greatly by the fact that they have been implemented using computer modeling languages. Implementations are shown in Chapter 7, which allows solutions to be found using a computer.

A reasonable question is: why write the book now? It is a combination of opportunities that have recently become available. The availability of modeling languages and computers that provides the opportunity to make practical use of the models that we develop. Meanwhile, software companies are providing software for optimized production planning in a supply chain. The opportunity to make use of such software gives rise to a need to understand some of the issues in computational models for optimized planning. This is best done by considering simple models and examples.

This book pursues a number of goals. Some of them are addressed directly such as the goal of developing useful models for production and distribution planning. Others accrue more as side-effects such as an understanding of the leverage that can be gained from abstract optimization models. We describe usable models that can be fed to fairly low cost, readily available software. However, many readers will be interested in these models not for direct implementation but as a means of understanding some of the issues in supply chain optimization for the purpose of using or assessing sophisticated, special purpose supply chain management software. We can also view the book either as a vehicle for understanding production planning within a supply chain via optimization modeling or for understanding optimization modeling by using production planning as an example. Both views are important.

These are things that need to be understood by managers and planners. This level of planning constitutes the interface between strategy and tactics. It is critical. However, it is often left to operational personnel or software vendors. This book aims to change that.

The book is also appropriate for a business school or industrial engineering course. Earlier drafts were used for a course in the Working Professional MBA program at the University of California at Davis, a business information systems course at the University of Technology in Braunschweig and the Karl Franzens University in Graz. Overall, the student population varied greatly from undergraduate business and engineering majors through professional managers from a wide range of backgrounds. The feedback from the students was a great help in preparing the final version of the book and we are grateful for it.

We would like to acknowledge the assistance of student Janin Oer who helped implementing the AMPL models shown in Chapter 7, Michael R. Bussieck of the GAMS development corporation who implemented the GAMS version of the models, Bjarni Kristjansson of Maximal Software who implemented the MPL versions, Greg Glockner and Veronique Blanchard of ILOG who implemented the OPL versions, and Susanne Heipcke of Dash Optimization who implemented the Xpress-Mosel models and students at the UC Davis Graduate School of Management as well as the staff at the University of Technology Braunschweig for proofreading.

Stefan Voß
David L. Woodruff

August 2002

Table of Contents

1. Introduction

Supply chain management rose to prominence as a major management issue in recent years. While the focus of managing a supply chain has undergone a drastic change as a result of improved information technology, production planning remains a critical issue. The ability to instantaneously exchange information along with increased computational power has enabled the use of sophisticated optimization software.

1.1 Supply Chains and Production Planning

Speculation about how long supply chains have existed degenerates quickly into an exercise in hyperbole. For our purposes, it suffices to say "longer than words." What is relatively recent is the discovery that optimization and a supply chain model of production and distribution is critical for competitive success. What is the supply chain model?

The idea is that the production of all but trivial goods and services can be thought of as a supply chain. The metaphor is very direct: a chain relies on interconnected links as does the production of goods and services. This abstract model is extremely useful. Notice that we did not mention organizational, factory or office boundaries, and did not distinguish between goods and services. At some level of abstraction goods and services are the same, and moves between machines in a factory share important characteristics with movement of goods between factories. Perhaps the most important thing from the perspective of improving management practices is recognition that the entire supply chain must be considered whether it is entirely within one factory or spans many organizations.

This is nearly the opposite of the *incentives philosophy* that gives rise to transfer pricing, management by accounting objectives and quantitative incentives for every division and plant manager as well as for every machine operator. In that philosophy there is an attempt to make every part of the supply chain seem like a separate organization to the extent possible. The idea is that if each participant in the supply chain has the proper incentives then they can myopically optimize their own performance and this may result in optimal system performance.

The trouble with the incentives philosophy is that it has proven to be nearly impossible to set up the proper incentives. The result has been managers and workers behaving in a way that is best described as gaming the system. They behave in ways that maximize the performance measures, but not necessarily in ways that help the organization as a whole or the customers of the supply chain.

This is not to say that incentives are not important. Often, the supply chain is composed of multiple arms-length players so contracts and other incentives must be structured in the best possible way. The creation of good contracts is also an optimization problem, but not one that concerns us here.

The value of optimized production planning within the supply chain is clear to many managers, particularly when large portions reside within one organization or the planning can be done in a coordinated way. The benefits from optimized planning for the entire supply chain that come as a result of coordinated planning have given rise to the concept of *partnering.* Organizations share information and planning functionality so that the benefits of a well run supply chain can be obtained. This shared information often comes from, and goes to, optimization models.

Our goal is to study optimization models that are general enough to be used for planning in a factory or across factory boundaries. However, we provide specific examples and computer implementations of the models. Before developing the models and their implementation details, we provide a brief introduction to concepts related to the modeling of optimization problems.

1.2 Optimization

Optimization is an activity that has existed even longer than supply chains have. What is relatively new is the explosion in supply chain optimization software. In order to truly understand this software and its proper role, we need to step back and understand optimization and how it can be applied to supply chains. The word *optimum* means "best." To be more precise, by *optimization* we mean the attempt to find the best possible solution to a problem.

In order to obtain understanding (or useful software, for that matter) we need to work at the appropriate level of abstraction. We build models but leave most of the data to be plugged in later. The models describe the structure of the problem. There is considerable leverage in this scheme because the same model can be used over and over again as the details change over time or it can be applied to a different organization by supplying the appropriate details. Furthermore, the models can be discussed in an abstract way that leads to better understanding of the underlying supply chain management issues.

In abstract optimization models, we use symbols to represent quantities. We begin with an overly simple example. Suppose that we have two products and we have to decide how much to produce next year. We might want to

maximize profit

subject to:

production of each product	not less than	strategic lower limits,
and total production	not greater than	capacity.

To produce a model suitable for abstract discussion, and also for exact solution, we can use the symbol x to represent the production quantity for the first product and y for the second. An assignment of specific values to x and y is referred to as a "solution," even if it is not the best possible solution. So if we say that x is 6 and y is 2, we have a solution that may or may not be a good solution or might not be possible to actually implement in the factory but we shall call it a solution.

Suppose that the profit from each product is known; we can use the symbols Q and R for these two numbers, respectively. Suppose further that we have values for the strategic lower limits on production of each of the two products and some capacity limitations on the total production quantity. We can model these in an abstract way as $x \geq A$, $y \geq B$, and $x + y \leq J$. We are using the symbols A, B, and J to give the respective limitations. How much should be produced? "It depends" is the correct answer, of course.

If actual numbers are supplied for A, B, J, Q, and R then we can quickly find the best possible values for x and y. Such models can be written to be read by people as well as computer software in the following fashion:

$$\text{maximize: } Q \times x + R \times y$$

subject to:

$$\begin{aligned} x &\geq A \\ y &\geq B \\ x + y &\leq J. \end{aligned}$$

Models written in this way offer a number of advantages:

1. It is useful to communicate the important structure of the problem without too many details getting in the way.
2. A related benefit is that it is easy to add or remove details. For example, we could add a separate capacity constraint for product one by adding a requirement that $x \leq K$. In this example, K would be used to represent the capacity limit.
3. These models can be fed directly to computer programs that can be set up to gather or maintain the data as well as compute the optimal solutions or good solutions when optimal solutions are too difficult to obtain.

In this book we will develop usable models that can be fed to general purpose (i.e., cost effective) optimization software. As an aside we note that these programs seldom require quite as much abstraction as we used in our simple model: there is no requirement that the variables have single character names, e.g., we could have used "TotalCapacity" instead of J. We use shorter names in the interest of saving space and highlighting similarities between models at a higher level.

1.3 Components of Supply Chain Management

In order to understand the scope of this book, we need to take a look at a list of the components of supply chain planning. Loosely speaking, we have organized the list so as to be decreasing in time horizon. That is to say that the first item, which is *strategy*, looks at the longest time horizons, while the last item considers the shortest. Each of these components can easily be the subject of an entire book of their own, so we include this list only with the modest goal of definition. The names that we have given to these components may have slightly different meanings depending on who is using them. That is one of the reasons we want to articulate them for our use here.

- *Strategy*
  We will not use the word "strategy" unless we are referring in some way to issues related to selecting and maintaining a sustainable competitive advantage and/or areas of core competency. Strategy planning involves determining how the organization will serve its potential markets and what its relationship will be with its competitors. In some ways, we can say that strategic planning seeks to answer the question "what" do we want to become (and "why") while tactical planning looks more at "how" we plan to do it.
- *Major Resources Capacity Planning*
  In this component we place decisions in the time frame of one to five years. To the extent that multiple organizations are involved in the supply chain, vendors and customers must select each other. In all cases, the parameters of compensation and evaluation must be established.
  Long term "make versus buy" decisions belong in this category. In the case of "make" we include decisions about production facilities along with decisions about entire personnel groups. This is intended to be distinct from capacity decisions made on a routine basis.
- *Tactical Production Planning*
  For our purposes, it will be useful to make a distinction between *scheduling* and *planning*. It is often said that schedules are for people who do the work and plans are for people who talk about work. In the planning activity we include aggregate planning and assignment of gross production quantities

to plants and/or lines. We also include routine capacity decisions such as assignment of work to a qualified subcontractor or activation of production lines. But we do not include sequencing decisions. Typical time frames for this type of activity are weekly time buckets extending to six or eighteen months. However, some organizations use monthly or daily planning time buckets.

- *Scheduling*
  This activity includes sequencing decisions and specification of production quantities at the level of individual part numbers. The time buckets used are sometimes the same as used for planning or sometimes down to the minute. This is the level at which procurement ship dates and due dates are established.
- *Execution and Feedback*
  This component includes the actual production, movement, shipping and delivery functions. We also include the mechanisms by which the results of these activities are communicated to the higher level activities. In many organizations these feedback mechanisms are woefully inadequate because they are designed primarily to satisfy accounting needs. One benefit of building optimization models is that it highlights the type of data that is needed to support effective long- and short-range planning.

This description of supply chain management is necessarily far too brief to capture important issues. However, it provides a context for the discussion that follows. Subsequent chapters deepen and clarify issues. Furthermore, we are able to provide a perspective on the relationships between models for optimized production planning in a supply chain and classical models of production planning.

1.4 Scope of this Book

This book addresses primarily the middle activity in the list: tactical production planning. There are some useful models and lessons for the neighboring activities in the list as well. We must be mindful of the need to integrate all of the activities in managing a supply chain as we develop models that assist in integrated and optimized planning and scheduling.

We will begin with models that are simple, yet produce valuable results. The understanding we get will help us to extend the models to include more detail. Eventually, we will find ourselves at the envelope of modern modeling and solution technology much of which is concerned with models that address issues of uncertainty and congestion. These advanced topics enable an understanding of the limits of commercially available supply chain management software. Also, they are very interesting and facilitate learning about the "physical laws" of production.

To lay the groundwork, we provide an introduction to modeling that is then used and expanded throughout the book. Although our primary interest is the models, we cannot ignore the methods for finding good or optimal solutions. Solution techniques are discussed in Chapter 8.

2. Optimization Modeling

A model captures the essential features of something without actually being the thing itself. Some models capture the shape and proportion of a physical object, but at a different scale and without the functionality. An example is a plastic model of a jet airliner. Some of these models are used as toys, but others are used to study air flows using a wind tunnel. When creating the model, some details have to be carefully reconstructed and others can be ignored entirely.

When building any model there is an art to including the important details and leaving out the rest. This is known as *abstraction*. There are two concepts that are needed; using symbols to represent general objects and selecting a sufficient level of detail to include in the abstraction. In this chapter we will acquire the skills needed to build the models and begin to develop a sense of the art of modeling.

2.1 Abstraction

We are all familiar with abstraction. We represent items and concepts using general symbols rather than something specific. A very simple, but illustrative example is the way that we discuss sales tax. We might tell someone that the sales tax is going to be "seven percent of whatever you spend." The words "whatever you spend" represent a number that is left unspecified. To be more general we might explain the concept of sales tax as "a percentage of whatever you spend." In this example, both numbers in the sales tax equation are left open. Almost everyone is comfortable with this type of abstraction.

There is a common set of shorthand that often is used with this type of abstraction. Let us exercise the shorthand. We might say that the sales tax, y, is a fraction, C, of whatever you spend on purchases, x. These symbols allow us to write the sales tax equation: $y = C \times x$. Since here we are using single characters for all of the variables and data, we (and most other authors) will use an even more condensed shorthand that assumes multiplication. We write $y = Cx$ to mean that the sales tax equals the tax rate times spending.

We can go a little further with this example. We can change the model slightly. Suppose we said to a friend from Nevada (where there is no sales tax) the following: "To compute the sales tax, you add up the cost of the items that you purchased then multiply the total times the tax rate." Notice that this statement is in some ways even more abstract in that it implies multiple items but does not specify how many.

Assign the symbol N to the number of items. One way we can use shorthand to write down this new equation is $y = C \times (x_1 + x_2 + \ldots + x_N)$. We have used the symbol $\ldots$ to mean "repeat the pattern" up to and including the ending value given. We have used subscripting to indicate that we want to represent the members of a list. We are saying that x_i represents the cost of the ith item that we purchased. A more compressed shorthand is available that uses the symbol $\sum$ to indicate summation over an indexed list. Using this notation the sales tax equation is

$$y = C \sum_{i=1}^{N} x_i$$

(or $y = C \sum_{i=1}^{N} x_i$).

Before proceeding with a careful development of the symbols that we need, let us pause to consider the philosophy of abstract models. Models that have the right level of abstraction can be very powerful. If there is not enough abstraction, we get bogged down in the details. With too much abstraction, we lack the details relevant to our analysis. The right amount of abstraction depends, of course, on our purpose.

If the goal of the model is to describe tax calculation for a purchase transaction the use of x_i for the purchase cost of item i seems appropriate. If we are interested in projecting tax revenues for the government from a variety of sources, we might want to model total sales tax revenues for year i as y_i. A model with considerably more detail would have $x_{j,k,l,m}$ the cost of item j purchased by person k on day l in year m for the purpose of projecting sales tax revenues by the government; but this is probably too much detail. Conversely, a model that represents total sales tax revenue for year i as y_i is so abstract that it has no value whatsoever for describing the tax calculation for an individual's purchase transaction.

In §1.3 we drew a distinction between planning and scheduling and stated that our main interest was in tactical planning. We can now recast this discussion using the terminology of abstraction that we have developed so far. We can say that for our purposes, planning models are more abstract than scheduling models. Scheduling models use finer granularity. The idea is that in planning we want to look a bit more at the forest than at the trees. We do not want to ignore the trees, but we do not want to be so focused on them that we miss the forest. This tired metaphor can be restated as: We want to use a model that is abstract enough to be manageable, yet contains enough detail to be realistic.

2.2 Symbols

Symbols enable condensed statements of abstract concepts. Once a reader is familiar with the symbols, they can often be easier to understand than words because natural languages are not particularly well suited for abstraction. Our purposes in this section are to introduce notation that will be useful throughout the book. Although some of the notation has already been introduced, more meaningful examples will be given in the following chapters. The reader should not be concerned if all of the notation does not seem intuitive at this point. A full description is given here so that this section can be used as a reference.

Shorthand is needed to write models for entire supply chains in a space small enough to be comprehended by someone. In exchange for the nuisance of learning these modeling conventions, we get very powerful and abstract models. The models can be used to get solutions to our planning problems. They can also be used as a basis for discussion and thoughtful consideration of planning, production, and distribution systems.

2.2.1 Variables, Data, Subscripts, and Math

When we want to specify a quantity that is to be determined in the model, we will usually use a lower case letter such as x. We refer to x as a variable. If we want to refer to a value that is to be supplied (or obtained from a database) as data, we will typically use an upper case letter such as C. Occasionally, we will use more than one letter for a single piece of data such as LS to indicate a lot size. Usually we will assume that appropriate units of measure are in use, but sometimes we will have to discuss this in the text.

Readers of different models have to be a little bit flexible since there is a limited number of letters. In one model C might be the sales tax rate, but in a different model it might be used to represent production capacity. Note also that in a model concerned with short term production planning the capacity would be data, but in a longer term model, the capacity might be a variable.

If the data or variables of interest can be organized into a list, we use subscripts. For example, we might use x_i to indicate production quantities for product i. When creating abstract models, it is often useful to assume that the products have been put in a list so it is reasonable to talk about the ith product. In similar fashion, we might assume that resources for production or shipping have been put in a list so we can represent the capacity of the jth resource as $C(j)$. Notice that we use parenthesis rather than subscripts for data that are in a list. This is done to help emphasize the difference between variables and data and to allow for data to depend on complicated expressions when necessary.

Often, we want to indicate that some variables are special. For example, we may want a variable to indicate that machine j has been set up for product i. Since either it has been or it has not been, this variable can take on only two values. For such variables, we will typically use Greek letters such as δ to help the reader of models identify special variables.

Sometimes the variables or data are naturally represented as lists of lists. The idea can be extended to lists of lists of lists and so on. In this book, we use multiple indexes that are separated by commas. For example we might use $C(j,t)$ to represent the production capacity for resource j in period number t and $x_{i,j,t}$ to represent production of product i using resource j in time period t. A list of numbers is sometimes called a *vector*. When we want to be less abstract, we sometimes just give the concrete values in a list using parenthesis to separate data from text. For example, if we have said that $C(j)$ is the capacity of resources j, we might give the actual values for the first three resources as $(12, 44, 56)$.

Arithmetic such as addition, subtraction, and division are shown in the usual ways. Multiplication is assumed when two symbols are side-by-side. For example, $D(j)x_j$ means that the jth element of the D list is to be multiplied by the jth element of the x variable list. When summation is to occur over a list or over two lists multiplied together, we will usually make use of a summation symbol, $\sum$. For example, if D and x are each lists of length greater than or equal to m we would indicate the sum of the product of the first m elements in the two lists as

$$\sum_{j=1}^{m} D(j)x_j.$$

Just to make this example concrete, suppose that $D = (4, 1, 2)$, $x = (2, 5, 1)$, and $m = 2$ then the sum will be $4 \times 2 + 1 \times 5 = 13$. If m happens to have a value that is the same as the length of the two lists, then this expression could be called the *product* of the two vectors. We use the notation x^2 to mean x times x.

We adopt the convention that multiplication and division are done before summation, which is done before explicit addition and subtraction. So the expression

$$\sum_{i=1}^{2} x_i + x_1 x_2$$

is 17 if $x_1 = 2$ and $x_2 = 5$. If the modeler needs a different order of operation, parenthesis can be used. Operations in parenthesis are done first, with the processing proceeding from the inner expressions to the outer as one would expect. There are better ways to write this, but we continue with our example where $x_1 = 2$ and $x_2 = 5$ and note that

$$\left(\sum_{i=1}^{2} [x_i + x_1]\right) x_2$$

is $[(2 + 2) + (5 + 2)] \times 5 = 55$.

In the interest of completeness, we need to mention the possibility of an *empty summation*. That is, the notation allows for a summation like $\sum_{i=3}^{t} x_i$ with t being data that need to be filled in. When t happens to be greater than or equal to 3 we know what to do. And in the event that t is smaller, then the summation becomes empty and we assume a result of 0.

2.2.2 Sets

The process of planning for production in a supply chain relies on the creation of many groupings. Products are grouped by family, shippers by mode, etc. These groupings can be given as *sets* and manipulated with the standard set operators. We will use calligraphic letters such as $\mathcal{L}$ to indicate sets. When we want to give a concrete example of the members of a set we will set brackets and a comma delimited list such as $\{23, 44, 6\}$.

Typically, we will want to collect sets of indexes and we also want to have indexed lists of sets. For example, we might indicate the set of product indexes that must use resource j by $\mathcal{U}(j)$. To be less abstract, suppose that parts 12, 13, and 21 are the indexes of the only parts that must use resource 17, then we would write $\mathcal{U}(17) = \{12, 13, 21\}$.

To indicate that an index variable i is to vary over all values in some set $\mathcal{A}$, we use the notation $i \in \mathcal{A}$. So the notation can be used to indicate summation over the members of the set $\mathcal{A}$. Continuing with the example using $\mathcal{U}$, suppose that x_i gives the amount of product i to be produced. We can give the total production by resource j as $y(j)$ and we can define it as

$$y(j) = \sum_{i \in \mathcal{U}(j)} x_i.$$

To be more concrete, we know that in this example

$$y(17) = x_{12} + x_{13} + x_{21}.$$

At the lowest level of abstraction, we might actually know of a decision that has been made concerning production quantities and we could compute the value of $y(17)$.

2.2.3 Objective Functions and Constraints

We have now assembled enough notation to begin to make some useful planning models. The most important step is to decide upon an objective. In many organizations, this is a very difficult step for the decision makers. Often, it is not possible to agree on a single objective, so the modeler must either merge multiple objectives into one or create multiple models.

It turns out that the very first decision that must be made is often one that has far reaching strategic and organizational implications. Should the supply chain minimize costs or maximize profits? As we shall see, they are not the same thing. We will return repeatedly to the implications of different objectives. However, for the moment, we will see how our notation can be used to write down an *objective function.* We call it a function because its value will depend on the values assigned to variables. The modeler does not assign values to variables; we rely on optimization software to do this.

As a practical matter, objective functions can be quite long. Let us begin with two simple ones. We will require that the user of our models lists the products so that they can be numbered, then we require that the cost for the ith item be given as $C(i)$. For the profit maximizing model, we will also need the revenue for each item $R(i)$. Assuming that there are N products, our objectives might be to

$$\text{minimize: } \sum_{i=1}^{N} C(i)x_i$$

or to

$$\text{maximize: } \sum_{i=1}^{N}(R(i) - C(i))x_i.$$

The expression $(R(i) - C(i))$ warrants comment. First, we note that parenthesis have been used to indicate that the subtraction is to be done before the multiplication. Second, clarity is almost always more important than efficiency, so we count on the software to handle the subtraction of data elements in $(R(i) - C(i))$. An alternative would be to have a data element called "profit" for each item which is equal to the revenue minus the cost of the item. Maybe we would use the symbol $P(i)$ for the data in this list. The advantage might seem to be that the subtraction can be done in advance, but this is really not much of an advantage since optimization packages are very good at this type of processing. The disadvantage is that the difference between the maximization and minimization objective functions would not be as obvious. We should almost always strive for clarity over efficiency since the time spent by the people who develop and interpret the models is usually more valuable than the computer time spent processing them.

Even these simple objective functions demonstrate the need for constraints in order to have the model make sense. One can see, for example, that costs are lower when nothing is produced than when something is produced. Actually, the cost objective function is minimized by producing negative quantities. An optimization program should make all of the x_i values the most negative number that can be represented by the computer. In similar fashion, profit is maximized by unlimited values for each x_i if revenues for the products are greater than costs. This may be mathematically the correct maximizer, but it is not useful.

This issue is so prevalent that many optimization software packages assume that variables must be constrained to be not less than zero unless otherwise noted. For production planning this avoids the absurdity of huge negative production values. But this highlights the major difficulty in constructing cost minimization models, which is that costs are minimized by producing nothing. This creates difficulties for production management that we will return to later. It also creates a need for ways to express constraints that we will address now.

In order to provide a fully specified model, we must add constraints on the decision variables. Constraints are given as inequalities (or sometimes equalities) that involve decision variables, data, and an indication of the indexes for which the constraints are to apply. For example, to constrain the cost minimization problem, we might at least want to add the requirement that $x_i \geq 0$ for $i = 1, \ldots, N$. Actually, a better thing to do would be to establish a data list of demand for each item. If we let $D(i)$ be the demand for product i, then a good constraint for the cost minimization objective function would be to require that production be greater than or equal to demand. The resulting problem would be to

$$\text{minimize: } \sum_{i=1}^{N} C(i)x_i \qquad \textbf{(MINC)}$$

subject to:

$$x_i \geq D(i) \qquad i = 1, \ldots, N.$$

There are a few things to notice. We have given the problem the name “MINC,” which is shown to the right of the objective function. Although we will always give the name with the objective function, we will always intend for it to refer to the entire problem with constraints. The next thing to notice is that we do not need a computer to solve this simple problem. Assuming that all of the $C(i)$ and $D(i)$ values are greater than zero the minimum cost production plan will be to produce exactly the quantity demanded of each item. The final thing to notice is our decision as modelers to require that production be greater than or equal to demand rather than equal to it. This is part of the art of modeling, and we will have more to say about it later.

Generally, equality constraints should be avoided unless they are clearly required for logical or physical reasons. No such reasons are present here; many organizations produce more than their demand. The reasons to use models with inequality constraints is that it more clearly indicates the intention of the modeler, it makes it so that the optimization software is more likely to be able to find a solution that satisfies all the constraints when additional constraints and objective function terms are added to the model to make it more realistic.

Another thing that some readers might notice is that we have not added any constraints of the form $x_i \geq 0$, that some modelers always add. Our feeling is that these constraints are not always a good idea, but they do offer some advantages. Consider the following problem:

$$\text{minimize:} \quad \sum_{i=1}^{N} C(i)x_i \qquad \textbf{(MINC0)}$$

subject to:

$$x_i \geq D(i) \qquad i = 1, \ldots, N$$
$$x_i \geq 0 \qquad i = 1, \ldots, N.$$

If our intention is that all of the $D(i)$ should be greater than zero, then constraining the variables to be greater than or equal to zero adds nothing but confusion because it is redundant. However, if we optimize the model when we fail to provide a value for the demand of some product, the production quantity will simply be set to zero. Whether or not this is good depends on whether or not it is an error if some products have no demand specified.

There is some terminology associated with constraints. We refer to the part before the relationship operator as the *left hand side* or the LHS. In **MINC0** the LHS is x_i for all constraints. In similar fashion we refer to $D(i)$ and 0 as the *right hand side* or RHS of their respective constraints. When we discuss a model, we often refer to a single clause such as

$$x_i \geq 0 \qquad i = 1, \ldots, N$$

using the plural "constraints." This is because the notation $i = 1, \ldots, N$ implies that there is a constraint on multiple members in a list (all members of the list, in this particular case).

Many people involved in production planning speak about "constraints" in daily work-related conversation. Sometimes they use the word very loosely and in other cases they use it in conjunction with various popular theories about constraints. One of our goals in this book is to introduce notation that links the concepts of constraints to emerging supply chain optimization software and to commercially available software for broader classes of optimization problems.

2.3 Finding Solutions

There is some value in just writing down an optimization model for supply chain planning. Many organizations send mixed messages about the goals of various planning groups and the exercise of creating a rigorous statement of the problem can highlight these issues and lead to important breakthroughs. However, often we want a computer to come up with a production plan that has the best possible value of the objective function. Companies that use special purpose supply chain optimization software make use of models that are "built-in" and other companies make use of optimization models that they create and maintain themselves. In either case, data must be supplied so that the software can search for a solution.

The process of finding a best possible solution is usually referred to as *optimization* and the software for this purpose is referred to as a *solver*. In everyday language we often use the word "optimization" for the process of improving solutions.

2.3.1 Data

The word "data" is used in many different ways. Often, any input to a computer is called data. Data *about* the model rather than data *for* the optimization are a big portion of the data that must be provided to major packages for supply chain optimization that are add-ons to larger accounting and enterprise management packages. This is why consultants are often required in order to install and tune the software. For companies that create their own models directly, data about the model are implicit in the models themselves.

In either case, some of the data *for* the models come from enterprise-wide databases and other data must be supplied specifically for the model. For example, data about the revenue from each product are often available from databases with little effort and data concerning costs are available with some effort. Optimization packages can read data from a variety of data sources. One of the tasks in setting up an actual optimization system is to specify data sources.

We are primarily concerned with creating models, but we cannot ignore the availability of reliable data. We must pause frequently to ask if one can reasonably obtain the data and what the effect will be if the data are uncertain. Even in the simple cost minimization model **MINC**, we face a data problem. Where will we obtain the demand data?

2.3.2 A Few Words About Uncertainty

Uncertainty makes the life of production planners much more difficult. The same can be said of supply chain optimization modelers. Much of the data that are needed for models cannot be known with certainty. For example, in the model **MINC** we rely on knowing the demands. It is often not possible to know the demands with certainty. One solution to this problem is to use the best guess available and plan accordingly, but it is often possible to do a better job than this. We will have more to say about uncertainty later, but for now we note that we can group methods of dealing with uncertainty in our models into two broad categories: 1) we can decompose the problem and deal with uncertainty separately or 2) we can explicitly include uncertainty in our models.

The first choice is currently more popular. When viewed from the perspective of the models that we develop, one can make use of a variety of formulas and algorithms to determine appropriate data for constraints that take into account the uncertainty. One can refer to this as decomposition of the uncertain and certain part. To use different language for the same thing, we might say that we decompose our problems into the *random* and the *deterministic* part. Some people use the word *stochastic* in place of "random."

An example of decomposition would involve using a model that suggests an optimal value of demand to use as a target. The result of this calculation would then be used as $D(i)$ in our calculations. These demand models can be simply forecasting models of varying sophistication such as those provided in commercial supply chain optimization software. They could also be models that are part of the academic research that attempt to consider cost structures as well. In the latter case, neither the models nor their inclusion in our models are trivial.

The second possibility is to incorporate the uncertainty in the modeling process. The most straightforward way to do this is by specifying *scenarios.* A complete set of data is specified for each scenario. In the **MINC** example, we would specify a full set of values for C and D for each scenario. After solving the problem for each of the scenarios, the decision maker can decide how to best hedge against uncertainty. A more sophisticated model, which is much more difficult to solve, involves associating a *probability* or a *likelihood* with each scenario. Some solvers can take this information into account when producing solutions. Complete details of all methods of dealing with uncertainty would fill an encyclopedia. However we address a few of the methods in later chapters.

2.3.3 Solvers and Model Structure

It was mentioned earlier that we typically want to create models that express our needs and are easy to read rather than worrying about what the computer will do with the model. We hedged a bit by using words like "usually" and "probably." This is because some things make the models much harder to solve optimally so we would want to put them in the model only if they are very important. These issues will be discussed as they arise, but we mention a few important ones here as illustrations.

Before worrying about whether or not a model can be solved optimally, we have to decide whether we care about optimality. In many situations, a solution that satisfies the constraints is good enough. By definition, we would always prefer a better value of the objective function, but having the best may not always be important. In other cases, the objective is of considerable importance and we would prefer to have an optimal value if possible. In either case, we would not want to do anything to make the solution harder to obtain without a good reason. For the moment we mention two things that can make obtaining an optimal solution much more difficult: integers and non-linearities.

It is often sensible to require some variables to take on integer values. For example, if the product that we produce is surgeries, it does not make sense to talk about producing a partial organ transplant. Nor does it make sense to allow for part of a flight attendant to travel between two cities.

Usually, the greater the number of variables that must be integers, the harder the problem is to solve. Consequently, although it also does not make sense to schedule the completion of a fraction of a 10mm bolt, we would typically not want to constrain the quantity of bolts scheduled to be an integer. This is because it will often come out to be an integer anyway and because even if it does not it will typically make no difference if it is rounded to an integer after the computer has reported a solution.

If there are a large number of integers inherent to the nature of the problem it is often preferable to make use of special solvers that are addressed later. For the time being, we will require integers only when it is critical to the model. Usually, we will want a variable to be constrained to be either a zero or a one, which we will denote with a constraint of the form $x \in \{0, 1\}$. If we need integers in some other range, say between J and K, we will use a constraint like $x \in \{J, \ldots, K\}$. A constraint enforcing a variable to be an integer is also called integrality constraint.

Nothing that has anything to do with planning is truly linear, but accounting is easier and so is optimization if we pretend that there is linearity. We will restrict ourselves to expressions where variables are not multiplied by other variables and where neither powers and roots of variables nor any other fancy functions are used. Such models are linear. In other words, expressions like x^2, x/y and $\sqrt{x}$ can not be part of linear objective functions and constraints provided that x and y are variables. Some kinds of non-linearities cause a lot less trouble than other kinds, but we will introduce non-linearities with great caution. For one thing, most production planning models are only an approximation anyway, so linear approximations are often good enough.

Non-linearities in the variables can make solutions harder to obtain, but non-linear expressions in the data do not cause any trouble. Sometimes it is useful to use expressions where roots of data are used particularly to help with conversions of units of measurement. In other cases, products and fractions involving data elements are used in modeling to make the model easier to understand. The non-linear expressions in the data can be evaluated when the values are provided and before the computer passes the problem to the solver software.

2.4 Implementing the Models in this Book

The models developed in this book can be fed almost directly to commercially available software. Nevertheless, it is important to keep some basic folk wisdom in mind when doing so:

- Model simple – think complicated
- Start small and extend
- Divide and conquer – try to avoid mega-models

- Try to use similarities, analogies, and metaphors
- Drive the data through the model and not vice versa

The models for supply chain management typically need extension and customization, but it is a good idea to begin with the models as given here. Models should be extended incrementally so that they can be debugged. Ultimately, it may become necessary to reduce the computational effort associated with solving the models. This may be accomplished, e.g., by using aggregation techniques such as those described in §6.6. A more immediately apparent issue when implementing the models is the need to add information about how to handle the last periods in the planning horizon. This is discussed in §6.3. Chapter 7 contains implementations using popular modeling languages.

One final word about problems and objective functions. As a convention we may assume, at least for a while, that all models and all objective functions in this book will be of the type "minimize" if not stated otherwise, to make the exposition clearer. If concepts are understood for minimization it is easy to generalize to maximization.

3. Starting with an mrp Model

Rather than creating a model from scratch, we begin with a venerable model called *materials requirements planning.* The model is often referred to as mrp with lower case letters (or sometimes "little-mrp") to make clear the distinction between mrp and MRP II. We will look at MRP II later.

The mrp model comes from a production planning perspective rather than an optimization perspective, but after we understand how it works, we can create an optimization model that corresponds to it. This model is a useful starting point for further modeling. So we will first understand the mrp model as it was originally given. It is a very practical model, so we will introduce an example early on. One does not need an optimization model for mrp, but we will use our model to further understand the limitations of mrp and we will use it as a basis for more sophisticated models.

The mrp model uses a lot of data about items and components. The term *Stock Keeping Unit* (SKU) is used to refer to items that are sold as well as components and components of components, etc. For each SKU we need to know

- the *lead time*, which is an estimate of the time between the release of an order to the shop floor or to a supplier and the receipt of the items;
- if there is a minimum production quantity (referred to as a *minimum lot size* for items that are manufactured in-house) or if there is a minimum order quantity for purchased items;
- the current inventory levels (for simplicity we include items scheduled for receipt during runs of mrp in earlier periods);
- components needed, which is often referred to as a bill of materials (BOM).

This list of data seems short, but can be very hard to obtain and maintain. The fact that mrp requires production personnel to provide and maintain these data is one of the reasons that mrp is often very popular with accountants.

3.1 An Example

We use a very small example to illustrate the notation. Suppose that there is a single end item with SKU AJ8172 that has a bill of materials as shown in Fig-

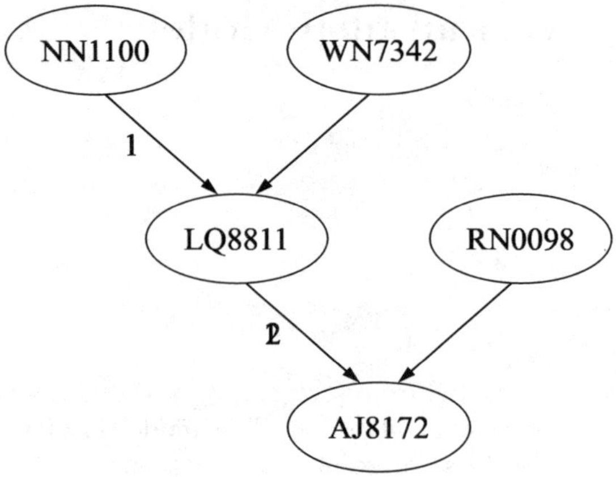

Fig. 3.1. BOM for a Simple Example

ure 3.1 and the properties given in Figure 3.2. In this figure, the components that are ordered from outside suppliers (e.g., RN0098) do not have a list of their components given since the mrp model typically stops at organizational boundaries.

While Figure 3.1 has a single end item, Figure 3.3 shows a slightly larger example that emphasizes that the product structure for mrp can be fairly general. In this example, there are two end items: AJ8172 and TR1777. They both use the component RN0098. The assembly of AJ8172 requires 1 of them and TR1777 requires three items of RN0098. We will generally make use of the simpler example, but the reader should bear in mind that mrp is intended for large numbers of SKUs with potentially many shared components or subassemblies.

3.2 mrp Mechanics

Materials Requirements Planning was defined operationally. Although it was invented before the term Just-in-Time (JIT) was popularized, in some sense mrp is a just-in-time system, even if it is not generally considered to be a "JIT" system. Production is planned to be done as late as possible but no later.

We make use of the so-called "low level" coding provided by mrp packages. This is an ordering of the parts such that the list begins with end-items and no item appears in the list before an item that contains it as a component. We assume that the parts are sorted in low level code order. We then proceed through the parts and for each one, we anticipate the need for lots to be ordered as inventories are depleted. Once we know when lots will be needed, we can subtract the lead time to determine when the order must be

1. AJ8172:
   Production Lead Time: 2 days
   Minimum Lot Size: 100
   Components: 2 LQ8811, 1 RN0098
   Initial Inventory: 90
2. LQ8811:
   Production Lead Time: 3 days
   Minimum Lot Size: 400
   Components: 1 NN1100, 1 WN7342
   Initial Inventory: 300
3. RN0098:
   Order Lead Time: 4 days
   Minimum Order Quantity: 100
   Components: N/A
   Initial Inventory: 100
4. NN1100:
   Order Lead Time: 1 day
   Minimum Order Quantity: 1
   Components: N/A
   Initial Inventory: 0
5. WN7342:
   Order Lead Time: 12 days
   Minimum Order Quantity: 1000
   Components: N/A
   Initial Inventory: 900

Fig. 3.2. Data for a Simple Example

released to the floor or the vendor. These release dates cause additions to the demand for component parts. When we eventually process the component SKUs, plans will be made to meet the accumulated demands for them, thus creating demands for their components and so on. Time must be broken into *buckets* such as days or weeks in an mrp model.

We can think of mrp logic as a plan for sequential toppling of a line of dominoes. If you know when you want the last domino to fall and you know the time for each of the dominoes to knock down the next one, you can calculate when the first domino should fall in order to achieve the desired result. The same logic underpins mrp.

Continuing with the example in Figure 3.2, suppose that the demand for AJ8172 in the next eight periods is 20, 30, 10, 20, 30, 20, 30, and 40. This results in the mrp production/inventory plan for AJ8172 given in Figure 3.4. We plan for receipts in the period when the inventory would be depleted without them. We then subtract the lead time in order to determine the last possible release date. Since we begin with an inventory of 90, the inventory would be depleted in period 5, so we plan to receive enough in period 5 to

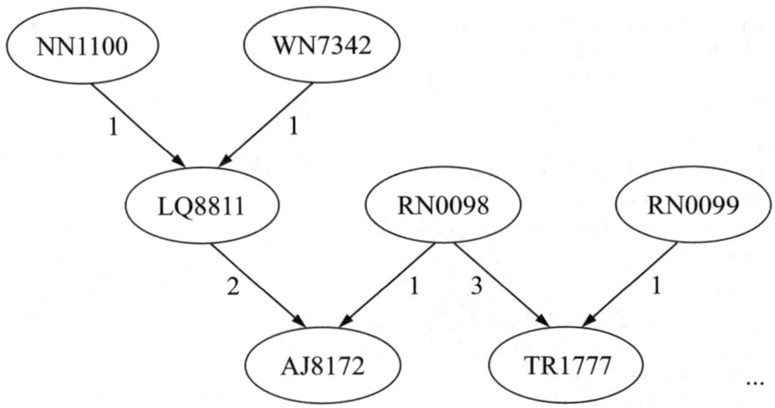

Fig. 3.3. BOM for a Slightly Bigger Example

avoid a shortage. In order to receive items in period 5, the order must be placed in period 3 because the lead time is 2. Since the minimum lot size is 100, the order is placed for that quantity.

	Day							
AJ8172	1	2	3	4	5	6	7	8
Demand	20	30	10	20	30	20	30	40
Inventory Plan (90)	70	40	30	10	80	60	30	90
Planned Receipts					100			100
Planned Releases			100			100		

Fig. 3.4. mrp Plan for AJ8172 Using Data from Figure 3.2

The planned releases of orders for AJ8172 creates demand for its components. There will be a need for 200 LQ8811s and 100 RN0098s in day 3 and day 6. This is shown in Figure 3.5. The initial inventory of WN7342 is adequate for the eight day planning horizon.

3.3 mrp Data

Central to mrp is the BOM, which gives the components that are combined to make other components and, ultimately, each end item. There are many ways to describe and display a BOM. Since we are interested in abstract models, we use the data $R(i, j)$ to give the number of SKU i's directly needed to make

	Day							
LQ8811	1	2	3	4	5	6	7	8
Demand			200			200		
Inventory Plan (300)	300	300	100	100	100	300	300	300
Planned Receipts						400		
Planned Releases			400					

	Day							
RN0098	1	2	3	4	5	6	7	8
Demand			100			100		
Inventory Plan (100)	100	100	0	0	0	0	0	0
Planned Receipts						100		
Planned Releases		100						

Fig. 3.5. mrp Plans for LQ8811 and RN0098 as a Result of Figure 3.4 Plan

one SKU j. This notation presupposes that the SKUs are in a numbered list such as Figure 3.2 so that the words "SKU i" have meaning. Any numbering of the parts can be used in the optimization models, but we use a low-level-coding to be somewhat consistent with mrp software. We let P represent the number of SKUs in the list. We use the numbering in Figure 3.2, which assigns a 1 to AJ8172, a 2 to LQ8811, and so on. Hence, the value of $R(2,1)$ is 2. The values of $R(3,1)$, $R(4,2)$, and $R(5,2)$ are 1. All others (e.g., $R(4,1)$, etc.) are zero.

Remember that time must be broken into "buckets" in an mrp model. The buckets are typically weeks or days. In our simple example we use days and we use T to represent the number of days in our planning model. In a typical mrp application T might be on the order of a few months or even a few years. Since T gives the last time bucket that we consider in the planning process, it is sometimes referred to as the *planning horizon*.

We use the notation $LT(i)$ to indicate the number of time buckets that one can expect between issuing an order for production or shipment of SKU i and receiving it. At this point, we are intentionally a little bit vague about the exact meaning. For the example data, $LT(1)=2$, $LT(2)=3$, etc.

Lot sizes are a standard part of mrp. Later, we will advocate elimination of most lot sizes as externally supplied data. But for now we want to remain consistent with common mrp practice, so we let $LS(i)$ be the minimum lot size for SKU i. In the example data, $LS(2) = 400$. For SKUs such as NN1100 that can be ordered in any quantity, LS should be given as 1. Other lot sizing conventions are possible, such as requiring that production be in multiples of a lot size. However, our models use a minimum production lot, which is satisfactory for avoiding unduly small production quantities for an SKU.

Our final pieces of data are the external demands for item i in period t, which is given as $D(i,t)$. Demands are clearly needed for end items. Of course, in many situations there are external demands for components as well because they are distributed to the maintenance organization, sold as replacement parts, or perhaps sold to competitors. This set of external demands is often called the *master production schedule.*

The data needed for an optimized mrp model are summarized in Table 3.1. A large number, referred to as M, is needed to force the computer to make some-or-none decisions that are needed to enforce the minimum lot sizes. This can be any number, provided it is larger than any possible production quantity. To avoid excessive roundoff error, one should try to use a number for M that is not more than, say, a hundred or a thousands times larger.

P	Number of SKUs
T	Number of time buckets (i.e., the planning horizon)
$LT(i)$	Lead time for SKU i
$R(i,j)$	Number of i's needed to make one j
$D(i,t)$	External demand for i in period t
$I(i,0)$	Beginning inventory of SKU i
$LS(i)$	Minimum lot size for SKU i
M	A large number

Table 3.1. Data for the mrp Formulation

3.4 mrp Optimization Formulation

An optimization model is not needed to use mrp, but we can create one and then extend it. In other words, our goal is to create an optimization problem that matches mrp not for its own sake but to get started with models that match classic planning systems. Using this model as a starting point, it is easy to go on to more sophisticated models.

We really have only one decision variable, $x_{i,t}$, which is the quantity of SKU i to start or order in period t. In order to enforce lot sizing rules, we need something that indicates production of an SKU in period t. What is needed is an *indicator variable.* This is a somewhat advanced topic, but we need to tackle it now in order to model classic mrp.

Note there are only two things in optimization models: data and variables. Something that indicates that there will be production in period t is clearly not data. We create a variable, $\delta_{i,t}$, that will be one if any of SKU i will be started in period t. We have to put in constraints to force the computer to make this variable behave this way. At first glance, this variable is redundant with $x_{i,t}$, but not equivalent. We will see that it serves a different role.

The following constraints must hold for all $i = 1, \ldots, P$ and $t = 1, \ldots, T$.

- Demand and materials requirement:

$$\sum_{\tau=1}^{t-LT(i)} x_{i,\tau} + I(i,0) - \sum_{\tau=1}^{t} \left(D(i,\tau) + \sum_{j=1}^{P} R(i,j)x_{j,\tau} \right) \geq 0$$

- Lot size requirement:

$$x_{i,t} \geq \delta_{i,t} LS(i)$$

- Modeling constraint for production indicator:

$$\delta_{i,t} \geq \frac{x_{i,t}}{M}$$

- Integer constraint for production indicator:

$$\delta_{i,t} \in \{0, 1\}$$

- Non-negative production:

$$x_{i,t} \geq 0$$

Fig. 3.6. Constraints for mrp

Figure 3.6 gives the constraints needed for a model of mrp. These constraints are fairly complicated, but they provide a nearly complete description of mrp. The first constraint requires that the sum of initial inventory and production up to each period has to be at least equal to the total of external demand and demand for assemblies that uses the SKU. The summation is to $t - LT(i)$ for each period (there will be one constraint for each value of t) because of work that must be started LT periods before it can be used to satisfy demand. The product $R(i,j)x_{j,\tau}$ anticipates the demand for SKU i that results when it is a component of SKU j. This product will turn out to be zero for a lot of i, j combinations, but that does not present any special difficulty for a computer.

The demand and materials constraint in Figure 3.6 could have been written

$$\sum_{\tau=1}^{t-LT(i)} x_{i,\tau} + I(i,0) \geq \sum_{\tau=1}^{t} \left(D(i,\tau) + \sum_{j=1}^{P} R(i,j)x_{j,\tau} \right).$$

The use of algebra to rearrange the terms of a constraint is entirely a matter of taste. There is no effect on the solution or the computational effort. We chose to put a zero on the RHS to emphasize that mrp requirements are all firm. The main point is that the term

$$\sum_{\tau=1}^{t-LT(i)} x_{i,\tau}$$

captures the production that will be completed up to time t, while the term

$$\sum_{\tau=1}^{t}\left(D(i,\tau)+\sum_{j=1}^{P}R(i,j)x_{j,\tau}\right)$$

is the total demand that will have occured up to the same time period.

The lot size constraint specifies that if there is any production of a SKU during a period it must be at least as much as the minimum lot size. The modeling constraint for the production indicator forces it to take a value greater than zero if there is production for the SKU in the period, which the integer constraint forces it to be either zero or one. Many modeling languages eliminate the need for these two constraints by allowing for *semi-continuous* variables, which must be either zero or above a threshold; see §8.3.1 for more information. The final constraint forces the computer to pick only production values that are not negative.

All we need in order to use an optimization package to accomplish mrp is an objective function. The objective in mrp is to make things as late as possible but no later. So one possible objective is to minimize

$$\sum_{i=1}^{P}\sum_{t=1}^{T}(T-t)x_{i,t}$$

that will result in things being made "as late as possible" and we count on the constraints to enforce the "but no later" part of the mrp model. There are better objectives than this, some of them will be described later. However, for now our goal is to mimic mrp. Stated in classic optimization form, the mrp problem is given in Figure 3.7.

In order to illustrate that these models can be implemented almost directly using a modeling language and make the notion of a modeling language concrete, Chapter 7 is provided. In that chapter, we demonstrate how some popular modeling languages can be used to implement this simple model **mrp**. We also show how the example data can be entered and an optimal solution obtained.

3.5 Discussion of mrp

Notice that we do not require integer valued production quantities. This particular model results in integer valued production quantities, provided that the demands and minimum lot sizes are integers. After we extend the model, we could worry that it might be possible for the computer to report that the optimal plan involves producing 123.3 of some product in a period and this might not make sense for that particular product. However, this model is intended to be a plan and not a schedule. Furthermore, unless the

Minimize:

$$\sum_{i=1}^{P}\sum_{t=1}^{T}(T-t)x_{i,t} \qquad \textbf{(mrp)}$$

subject to:

$$\sum_{\tau=1}^{t-LT(i)} x_{i,\tau} + I(i,0) - \sum_{\tau=1}^{t}\left(D(i,\tau) + \sum_{j=1}^{P} R(i,j)x_{j,\tau}\right) \geq 0$$
$$i=1,\ldots,P,\; t=1,\ldots,T$$

$$x_{i,t} - \delta_{i,t}LS(i) \geq 0 \qquad i=1,\ldots,P,\; t=1,\ldots,T$$

$$\delta_{i,t} - \frac{x_{i,t}}{M} \geq 0 \qquad i=1,\ldots,P,\; t=1,\ldots,T$$

$$\delta_{i,t} \in \{0,1\} \qquad i=1,\ldots,P,\; t=1,\ldots,T$$

$$x_{i,t} \geq 0 \qquad i=1,\ldots,P,\; t=1,\ldots,T$$

Fig. 3.7. mrp Model

production quantities are very small, the production quantity can typically be rounded to the nearest integer producing errors that are much smaller than the estimation errors involved in the data for the problem.

Classic versions of mrp, as well as our optimization formulation, are intended only for certain types of bills of material. Bills of material where multiple SKUs are combined to make a new SKU work well. This is the case with many things such as computers and cars. Products where one item is used to produce multiple items are referred to as *divergent* BOMs. For divergent portions of the BOM, the entries in $R(i,j)$ will be fractional. For example, suppose SKU TB4-16 is a sixteen foot board and TB4-8 is an eight foot board. In this example R(TB4-16, TB4-8) will be one half. More complicated situations such as cycles in the BOM, require modification to the basic scheme. Such modifications are beyond our scope.

We close our discussion of mrp by considering some of its troubles along with some of its virtues. Computerized planning systems based on mrp have been in use for decades and mrp logic remains at the heart of the production planning module of many modern *Enterprise Resources Planning* (ERP) systems. Any system in broad use is sure to have some troubles and virtues.

3.5.1 Troubles

Of course, no production planning model can be perfect, but there are a number of well-known and very severe problems with mrp as we have described it. Perhaps the three most serious problems are

- the actual time to complete an order is usually a function of congestion rather than of the SKU,

- lot sizing can cause *nervousness*,
- there are no capacity constraints.

It turns out that these problems are related. The lack of capacity constraints results in a need for lot sizing and exacerbates the variable lead time problem considerably.

The underlying model for mrp is based on common thinking among data processing professionals in the early days of the application of computers to business problems. Data are collected into a database and then processed. The trouble with lead times is that they are given as static data. However, the time from issuance of an order to completion depends mainly on what work has to be done before the order "gets to the front of the line" of orders that await processing. Lead times are often weeks when the actual production time is hours.

One reason that lead times have to be much longer than production times is to account for machine failures. But even if the capacity is not overutilized and if the production resources are reliable, lead times can be variable due to waiting lines that form in front of bottleneck resources. This issue can be important but it is difficult to deal with, so it must be deferred to a later research oriented chapter. Apart from the (considerable) mitigation due to using capacity constraints, discussion the problem of variable lead times is deferred to Chapter 9.1.

A major part of the reason that lead times must be so long is to guard against periods when the resources are overbooked. This can be mitigated considerably by the use of capacity constraints, which are not included in mrp. As we shall see, MRP II was developed as a partial solution to this problem. We will develop optimization models that address capacity constraints much more effectively.

Nervousness is a phenomenon where small changes in demand result in large changes in production plans due to lot sizing rules. Consider the example given in §3.2. If the company received an order for ten more AJ8172's in period four, it would cause production to be shifted earlier and production for the entire lot of 400 LQ8811's would be shifted one period earlier as well. The dynamics of nervousness makes it demoralizing for production workers and as a result they often ignore the production plans produced by mrp systems. Who can blame them? First the production plan given in §3.2 is released to the floor, then a modest size order causes large changes in the schedule; meanwhile order cancellations can have a similar effect.

One response to this problem is to produce a "frozen zone" for end items that forbids changes in the schedule for some number of near-term time buckets. This seldom works. The reason is that customers do not care about mrp induced nervousness. They demand flexibility. As a result, orders get changed whether the master production schedule reflects it or not. In the worst case (a common case, unfortunately) expeditors manipulate inventories and production schedules to respond to customer needs. Once this practice begins, it is hard to stop. The mrp system assumptions are no longer valid at all, since components originally produced for one end item are used in another. Eventually, day-to-day scheduling is essentially done by expeditors and the mrp system is reduced to a raw materials procurement aid.

A major reason to use large lot sizes, or lot sizing rules at all, is to ensure that not too much productive capacity is used to changeover from one SKU to another. Lot sizes are at best a blunt instrument for accomplishing this. In reality, one cannot know how big the lot sizes need to be until the production schedule is complete. For resources that do not happen to be capacity constrained in a time period, the addition of more changeovers will not adversely affect throughput, so smaller, more flexible lots can be used. Conversely, for resources that are capacity constrained, a delicate tradeoff is needed between the use of small lots to provide flexibility in meeting customer needs and the use of large lots to maximize throughput. Setting lot sizing rules *a priori* is hardly delicate.

The solution is to simultaneously create production schedules and determine lot sizes that respect capacity limitations. This solution shows the final problem with mrp: there is nothing in mrp to guarantee production schedules that can actually be executed. That is to say, there is an excellent chance the production plans for many SKUs far exceed the capacity of the resources used to create them.

3.5.2 Virtues

Having said all that, we can say that an mrp model can still be very useful. For one thing, it is usually much better than no planning model at all. This is particularly true in industries with changing demand patterns where standard orders cannot be used. An mrp model can provide a good starting point for planning and for the ordering of raw materials.

The materials requirements estimates provided by an mrp system can be useful to the purchasing department because they provide an "earliest-case" estimate of requirements. To the extent that the plan exceeds available capacity, the actual production will take place later than the mrp plan. The idea of buying materials based on an "earliest-case" estimate is not consistent with modern notions of "just-in-time," but it is better than being late. Furthermore, for commonly used low cost raw materials, the mrp plan can provide essential information for purchasing.

Another reason to consider the use of an mrp model is the same reason that causes us to begin with it in this book. It is a simple model that can be the starting point for more sophisticated models. If you plan to use an mrp model in production and there are minimum lot size requirement, you should see §8.3.1 for a method of streamlining the model.

4. Extending to an MRP II Model

MRP II was inspired by shortcomings in mrp, and as a result the data processing orientation is preserved in MRP II. As was the case with mrp, we first explain the concepts behind MRP II, then we develop an optimization model to mimic and improve its behavior. After we have this model in hand, we extend it to produce a model that can give us production plans that trade off alternative capacity uses, holding inventory and tardiness in an optimized way. The letters MRP in MRP II stand for *Manufacturing Resources Planning* to make it clear that resources are considered in addition to materials as in mrp. The word "resource" is used to emphasize that any type of productive capability can be considered, not just machines. The Roman number II is intended to make it clear that it is an extension to materials requirements planning (mrp).

4.1 MRP II Mechanics

There are a number of well-known deficiencies in the model that underlies mrp. Potentially the most severe is the fact that it ignores capacity. To discuss this issue it is useful to remember that we are making a distinction between planning and scheduling as we described in §1.3. Although we have introduced it as a planning tool, mrp is also often used as a scheduling tool as well. A severe problem is that there is no guarantee that there will be enough capacity to actually carry out the plan produced by mrp. In fact, for capacity constrained production systems, it is seldom possible to implement an mrp plan as a schedule. This is debilitating when mrp is used as a scheduling tool, but is also bad for mrp as a planning tool because the absence of capacity considerations can make the plans so unrealistic that they are not useful.

The data processing professionals who were developing and selling mrp software in its early years recognized this deficiency and MRP II was developed in response to it. The database for mrp is extended to include routing and capacity information. Each production resource is entered into the database along with its maximum production during a time bucket. We will refer to the maximum production by a resource during a time bucket as its *capacity.* The list of resources used to produce a particular SKU is known as the *routing* for the SKU.

With this information the data processing specified by MRP II can be carried out. The processing begins by executing mrp to determine a production plan. Then, for each time bucket each SKU is "followed along its routing" and the utilization of each resource is updated. In the end, those resources whose capacity would be exceeded by the mrp plan are identified. The user can be given long and potentially confusing reports that dice and slice the following data for infeasibilities:

- resources,
- time buckets, and
- SKUs.

Reports are also generated for end-items that would use these "offending" SKUs and the time buckets in which the end-items would be produced.

The information concerning capacity infeasibilities can be used by the planner or some software to attempt to change the input data so that a feasible plan results. The most common method is to "repair" the master production schedule (i.e., change the timing of the external demand input data.) Such repairs are difficult, so one of our goals will be to use optimization models to produce feasible plans in the first place and eventually improve them later on. But before we do that, we provide the details of MRP II for the example given earlier. The example is too small to provide a realistic view of the issues associated with creating feasible plans in a real production setting, but it will be good enough for us to illustrate the mechanics of MRP II.

Before we can get started with a practical example, we need to deal with the issue of units of measurement. In a specific production facility capacity may be measured in hours, or tons, or pieces, or something else. Since we want to create abstract models, we will designate the capacity of every resource during every time bucket to be one. This will allow us to state resource utilizations as fractions in our models. We will represent the fraction of the capacity of resource k in one time bucket used by production of one SKU i as $U(i,k)$.

Suppose that we have only two resources for in-house production: HR-101 and MT-402. Further suppose that HR-101 has 80 hours of capacity per day and that MT-402 can produce 300 items per day. A capacity of 80 hours in one day would typically be achieved by a crew of 10 people (or "human resources"). There are only two items in Figure 3.2 that are produced in-house. Suppose that AJ8172 requires 10 minutes of HR-101 and that LQ8811 requires 5 minutes of HR-101 and also makes use of MT-402. Using common slang, we would say that LQ8811 is routed through both HR-101 and MT-402, while AJ8172 is routed only through HR-101. Note that even though the word "through" is used, it might be the case that HR-101 is a person or group of people who move from job to job. This routing is shown in an abstract way in Figure 4.1.

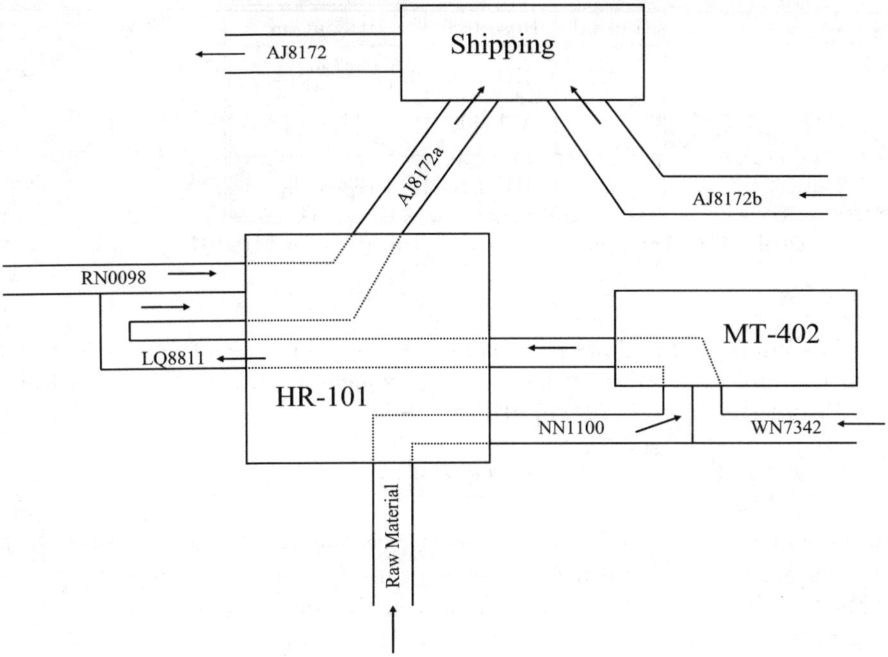

Fig. 4.1. Simple Routing Diagram

To use our notation, label HR-101 as resource 1 and MT-402 as resource 2. We then calculate that the utilization fraction of AJ8172 on HR-101 is

$$\frac{(10/60)\text{ hours}}{80\text{ hours}} = \frac{1}{480}.$$

Later, we will refer to this as $U(1,1)$. In similar fashion, we can compute the utilization fraction for one unit of LQ8811 of HR-101 and MT-402 as $\frac{(5/60)}{80} = \frac{1}{960}$ and $\frac{1}{300}$, respectively. We have summarized these utilization fractions in Table 4.1.

	Fraction Used By	
Resource	AJ8172	LQ8811
HR-101	1/480	1/960
MT-402		1/300

Table 4.1. Fraction of Capacity Utilized to Make Each SKU at Each Resource

Period	Resource	Utilization
3	HR-101	$\frac{100}{480} + \frac{400}{960} = \frac{5}{8}$
	MT-402	$\frac{400}{300} = \frac{4}{3}$
6	HR-101	$\frac{100}{480} = \frac{5}{24}$

Table 4.2. Anticipated Capacity Utilization for MRP II Example

This information allows us to compute the utilizations that would result from the mrp plan we developed in §3.2, which is given in Figures 3.4 and 3.5. Production of AJ8172 will utilize

$$100 \times \frac{1}{480}$$

of the capacity for HR-101 in periods three and six. In period three, production of LQ8811 will utilize $\frac{400}{960}$ of the capacity for HR-101 and $\frac{400}{300}$ of the capacity for MT-402. We have summarized the anticipated capacity utilization for the mrp production plan in Table 4.2.

We can see that the plan would not be possible because there is not enough capacity for MT-402 in period three. One solution is to move some of the production of LQ8811 to period two. Other possibilities include authorizing overtime for MT-402 or subcontracting some of the production of LQ8811.

It is not too hard to fix things with a small example, but when there are hundreds or thousands of SKUs interacting on many resources, it can be very difficult. In the next sections we describe ways to use optimization models and software to find good, and perhaps optimal, solutions based on the MRP II model.

4.2 MRP II Data and Constraints

As was the case with mrp, MRP II did not begin as an optimization model. However, we can create an optimization model that will mimic its behavior and do more. To be specific, we can mimic its intended behavior. That is, we can schedule things as late as possible without violating capacity constraints. The objective function for mrp is retained, but additional data are needed for constraints. The data needed to mimic MRP II are given in Table 4.3.

The data requirements are nearly the same as for mrp except that we have dropped the lot sizing information (for now) and added information about utilization. We use the same variables as for mrp, namely $x_{i,t}$, which is the quantity of SKU i to start or order in period t. We will not need $\delta_{i,t}$ unless we need to add information about changeovers. The major change from the

P	Number of SKUs
T	Number of time buckets
K	Number of resources
$I(i, 0)$	Beginning inventory of SKU i
$LT(i)$	Lead time for SKU i
$R(i, j)$	Amount of SKU i needed to make one j
$D(i, t)$	External demand for SKU i in period t
$U(i, k)$	Fraction of resource k needed to make one unit of SKU i

Table 4.3. Data for a Simple MRP II Formulation

mrp model is the addition of a capacity constraint. The MRP II constraints are as follows:

- Demand and materials requirement for all times t and all SKUs i:

$$\sum_{\tau=1}^{t-LT(i)} x_{i,\tau} + I(i,0) - \sum_{\tau=1}^{t} \left[D(i,\tau) + \sum_{j=1}^{P} R(i,j) x_{j,\tau} \right] \geq 0$$

- Constrain capacity for some (or all) resources k and times t:

$$\sum_{i=1}^{P} U(i,k) x_{i,t} \leq 1$$

- Non-negative production for all SKUs i and times t:

$$x_{i,t} \geq 0$$

Stated in classic optimization form, the MRP II problem is given in Figure 4.2.

For the purpose of discussion it is useful to isolate the two most important constraints and refer to them by name. We will refer to

$$\sum_{\tau=1}^{t-LT(i)} x_{i,\tau} + I(i,0) - \sum_{\tau=1}^{t} \left[D(i,\tau) + \sum_{j=1}^{P} R(i,j) x_{j,\tau} \right] \geq 0$$

as the *materials requirements constraint*, and we will call

$$\sum_{i=1}^{P} U(i,k) x_{i,t} \leq 1$$

the *capacity constraint*.

If there are minimum lot sizes for some SKUs, then we must add the following constraints for those SKUs, i:

Minimize:

$$\sum_{i=1}^{P} \sum_{t=1}^{T} (T-t)x_{i,t} \qquad \textbf{(MRPII)}$$

subject to:

$$\sum_{\tau=1}^{t-LT(i)} x_{i,\tau} + I(i,0) - \sum_{\tau=1}^{t} \left(D(i,\tau) + \sum_{j=1}^{P} R(i,j)x_{j,\tau} \right) \geq 0 \qquad i = 1,\ldots,P, \; t = 1,\ldots,T$$

$$\sum_{i=1}^{P} U(i,k)x_{i,t} \leq 1 \qquad t = 1,\ldots,T, \; k = 1,\ldots,K$$

$$x_{i,t} \geq 0 \qquad i = 1,\ldots,P, \; t = 1,\ldots,T$$

Fig. 4.2. MRPII Model

$$x_{i,t} - \delta_{i,t}LS(i) \geq 0 \qquad t = 1,\ldots,T$$

$$\delta_{i,t} - \frac{x_{i,t}}{M} \geq 0 \qquad t = 1,\ldots,T$$

$$\delta_{i,t} \in \{0,1\} \qquad t = 1,\ldots,T$$

where $LS(i)$ is the minimum lot size for SKU i and M is a large number as in the mrp model.

If capacities are expected to change for a resource, k, then the capacity constraints for these resources must be written as

$$\sum_{i=1}^{P} U(i,k,t)x_{i,t} \leq 1.$$

4.3 Discussion of MRP II

Using classic MRP II software, problem **MRPII** would not be solved directly. Instead, problem **mrp** would be solved and then the capacity constraint for the **MRPII** model would be checked. To be more specific, suppose the solution to the problem **mrp** was given as $X(i,t)$ for the production of SKU i to plan to start in time t. In other words, the result of solving problem **mrp** provides values for the decision variables. Once these values are known, they become data for subsequent processing so we use an upper case X to indicate the values that are given.

Given the **mrp** solution, those SKUs for which

$$\sum_{i=1}^{P} U(i,k)X(i,t) > 1$$

can be identified as those that would violate the capacity constraints. They would be the subject of reports and the production planner would change the data in an attempt to find a solution to mrp that was also feasible for **MRPII**. By "change the data" we mean that due dates, lot sizes, and capacity utilizations would be changed. Due dates for end items are typically adjusted to be at a later date. These data would then be given to the software that uses the new data to compute a new solution to **mrp** and then checks the constraints for **MRPII**.

This process is very hard work for the planners. Even though MRP II software provides numerous reports, it is often still not possible for the planners to produce a capacity feasible schedule given that they often have only a few hours to produce it. With a few dozen time buckets and a few thousand SKUs, the problem is just too big to be solved easily by people, even if assisted by software that can check the constraints and solve problem **mrp**.

An important thing to note is that the classic iterative MRP II solution process that we have described results in an implicit, rather than an explicit objective function. The actual objective function that is implied by the MRP II solution process is not easily determined because the solution is obtained by changing the data rather than finding the best solution given the best estimate of the data. As the planners change the data during the struggle to find a good, feasible solution the objective function implicit in the solution meanders without clearly articulated direction. The fact that some modern ERP software makes it easy to change the data (for example to increase lead times) emphasizes the importance of this point.

In spite of the severe difficulties, at the time of this writing MRP II logic is central to modern ERP systems. Many academics are of the opinion that MRP II should be done away with. If they mean that we should dump the solution methods, then we agree, but if they mean dump it completely, then they are hopelessly misguided. Regardless of how one articulates the methods, one should include the requirements constraint and the capacity constraint. They represent physical reality. The processing for MRP II is not normally explained by giving the constraints as we have shown them, but the constraints are there. They have to be. They have to be there in any other system as well. If the software does not include them explicitly or implicitly, then the planners will have to enforce them or else produce a plan that is very unrealistic and therefore not very valuable. Ultimately, the constraints are present in the production process.

Direct solution of the optimization model is a much better idea and this is the basis of much of the newest planning software that is sold as modules or add-ins for ERP systems. In practice, the problem is bigger and harder to solve than the simple **MRPII** model that we have presented. However, **MRPII** provides us with a good jumping off point for more sophisticated models because it mimics a widely used planning tool.

We can and will embed these constraints in a model that captures costs and constraints that are important to the manufacturing organization or the supply chain. By solving the optimization problem directly, we can include in the objective function guidance for solutions. We want to find solutions that correspond to the goals of the organization over and above merely satisfying the two constraints.

4.4 Changeover Modeling Considerations

The simple **MRPII** model will be the basis for many additional features. However, we might also want to remove features. For example, not all resources need to be modeled. Often, it is easy to see that some resources do not pose a problem. Such resources should simply be omitted from the model.

One feature that has been dropped from the **mrp** model in creating the **MRPII** model is lot sizes. Usually, the valid reason for minimum lot sizes is that a significant effort or cost is required to changeover for production, hence small lots might not be cost effective. Setting a fixed lot size is a crude response to this problem, so we will try to build models that take changeovers explicitly into account.

However, there are cases where minimum lot sizes are needed. For example, some chemical processes must be done in batches with a minimum size. If needed, the δ variables can be put in for the SKUs that require lot sizes along with the lot sizing constraints for those SKUs.

In many production environments, the proper modeling of capacity requires modeling of changeovers. By the word "changeover" we mean the effort required to switch a resource from the production of one SKU to the production of another. In fact, it is changeover avoidance that results in the need for lots that are larger than what would be needed to satisfy immediate customer demands. Changeover modeling can be quite involved.

The first thing that we will need is $\delta_{i,t}$, which will be equal to one if any of SKU i will be started in period t. This, in turn, requires that we include the constraints introduced for **mrp** to enforce the meaning of the δ variables.

- Modeling constraint for production indicator for all SKUs i and times t:

$$\delta_{i,t} \geq \frac{x_{i,t}}{M}$$

- Integer constraint for production indicator for all SKUs i and times t:

$$\delta_{i,t} \in \{0, 1\}$$

4.4.1 A Straightforward Modification

In a simple model of changeovers, we use the data given in Table 4.4. As usual we do not insist that these data be given for every resource or every

SKU. In particular, we would expect that W values will essentially be zero for most or all SKUs. If they are all zero, then of course there is no need to add them to the model. The idea is that when one changes from one SKU to another, some material can be destroyed (i.e., wasted.) A very common example is that when production for SKU j is begun, a few items of that SKU have to be destroyed for quality control testing or a few defective items are produced while the machine is adjusted. For example, suppose for SKU 32, 100 items are created during a changeover that have to be discarded. In this case $W(32, 32)$ would be 100. In more complex situations, only some of the components of SKU j are wasted during the changeover so we allow for a more general data element, $W(i, j)$, to facilitate this.

$S(i, k)$	Fraction of resource k used to changeover to SKU i
$W(i, j)$	Waste of SKU i to changeover to SKU j

Table 4.4. Data for Changeover Constraints

The following replacements for the requirements and capacity constraints are needed:

- Demand and materials requirement for all times t and all SKUs i:

$$\sum_{\tau=1}^{t-LT(i)} x_{i,\tau} + I(i,0) - \sum_{\tau=1}^{t} \left[D(i,\tau) + \sum_{j=1}^{P} (R(i,j)x_{j,\tau} + W(i,j)\delta_{j,\tau}) \right] \geq 0$$

- Constrain capacity for all resources k and times t:

$$\sum_{i=1}^{P} (U(i,k)x_{i,t} + S(i,k)\delta_{i,t}) \leq 1$$

4.4.2 Production that Spans Time Buckets

In other environments, the fact that an SKU is produced in a time period does not necessarily mean that there will be a changeover in the time period because the production run might span multiple time periods. If this is the case, the expression $W(i, j)\delta_{j,\tau}$ will overstate the requirements and the expression $S(i, k)\delta_{i,t}$ will overstate the capacity utilization. The proper correction depends somewhat on the situation. One solution is to introduce variables $\gamma_{i,k,t}$ that are usually zero, but take the value one if SKU i will be the last product produced on resource k in time bucket $t-1$ and the first produced in time bucket t. We can change the capacity constraint for some (or all) resources k and times t to be

$$\sum_{i=1}^{P} [U(i,k)x_{i,t} + S(i,k)(\delta_{i,t} - \gamma_{i,k,t})] \leq 1.$$

In order to force the γ variables to have the appropriate meaning we must require that they be either zero or one and we must add the following constraints for all t and for those k and i of interest:

$$\delta_{i,t-1} + \delta_{i,t} \geq 2\gamma_{i,k,t} \tag{4.1}$$

$$\gamma_{i,k,t}/M \leq U(i,k) \tag{4.2}$$

$$\sum_{i=1}^{P} \gamma_{i,k,t} \leq 1. \tag{4.3}$$

The constraints labeled (4.1) allow γ to be one for SKU i on resource k only if there is production of SKU i in both periods. Constraints (4.2) ensure that we only set γ to one for SKUs i that are to be routed to resource k, which is done mainly to avoid spurious values of γ that can be confusing when reading the solution. Constraints (4.3) ensure that at most one product can span the time boundary on a specific resource k. In order to have constraints (4.1) make sense, we must define $\delta_{i,0}$ as data, which is a departure from our usual naming conventions that use parenthesis for data indexes. In other words, $\delta_{i,0}$ is set to one for the SKUs that are scheduled for production in the last period before the first period for which the model applies. In many cases, the planning model begins with the "next" period so the $\delta_{i,0}$ values would reflect the schedule for the "current" time bucket.

The drawback to this scheme is that increases in the number of integer variables can dramatically increase the time needed for solving the problem. This can be mitigated somewhat by modifying the constraints and providing the variables only for resources k that are bottlenecks with significant changeover requirements. In other words, we do not create variables $\gamma_{i,k,t}$ for those k that do not have significant changeovers or that do have plenty of capacity in every period.

4.4.3 Parallel Machines

When there are a number of identical machines that are grouped together, it is possible for an SKU to be run on more than one machine at a time. This means that the amount of capacity consumed by changeovers depends on the number of machines used.

This presents yet another modeling dilemma. To model parallel machines properly, a large number of integer variables must be introduced. If there are many such resource groupings, and if they are potential bottlenecks and many of the parts use the bottlenecks, then special purpose sequencing and assignment algorithms must be used. This sort of situation is beyond the scope of this book. However, if there are only one or two resource groups with parallel machines that are important then it can be reasonable to add their descriptions to a linear model.

One way to model them exactly is to treat each machine as a separate resource and then every SKU that can use the set of parallel machines will have alternative routings. Modeling of alternative routings is discussed in §6.1. Unless the number of machines is fairly small, this may add too much complexity.

A computationally simpler solution is to use an approximation. We can model the changeover time with an approximation based on typical uses of the machines. For example, we might let the changeover time be a multiple of the single machine changeover time for some SKUs to anticipate that more than one machine will be changed over for those SKUs. A somewhat better approximation can be obtained by modeling the effect of quantity on the utilization rather than modeling the changeovers explicitly.

One can use historical or engineering data to fit a function that predicts the utilization (including changeovers) based on the quantity produced. Consider a situation where a group of machines is typically used to process one family of SKUs and there is only a minor changeover required between members of this family. However, the machine group might also process a few other SKUs requiring a significant changeover. Consider one such SKU, i, for a group of parallel machines that is resource k. If just one part is produced the utilization will be quite high because a changeover will be required in addition to the processing time. If the same machine can be used to produce up to 100 including a changeover, then the per part utilization (given as $U(i,k)$) will be a decreasing function of x up to 100 parts, then increasing sharply, then decreasing again. Modeling this sort of non-linear constraint is deferred to §8.3.

4.4.4 Sequence Dependent Changeovers

In some production environments, the changeover effort depends not only on the SKU that is to be produced, but on the previous part as well. The data for this can be modeled by changing the S data element to have an additional subscript. We can use $S(i,j,k)$ to indicate the capacity used to switch production on resource k from SKU i to SKU j. It is then necessary to model the production sequence.

In order to make use of linear models, we need to have indicator variables for the sequence. For example, we could add a subscript to the δ variables so that the production sequence is indicated. Under this scheme, the variable $\delta_{i,j,t}$ is equal to one if product i is produced immediately before product j in period t. This results in a set of rather messy constraints. There are a number of choices for writing the constraints that force creation of a sequence. We give one method here that relies on creation of a fictional SKU number 0 that serves as the first and last SKU in the sequence for each period. Presumably $S(0, j, k)$ and $S(i, 0, k)$ can have an average changeover utilization value for the affected resources. Clearly, if there are only a few SKUs produced in each time bucket, then this model will not be very accurate and an even more complicated model may be needed.

This adds many new constraints and there will be a dramatic increase in the number of variables. If there are P products and T time periods, then there would be more than TP^2 integer variables. For even a small company with 12 periods in the planning horizon and 100 SKUs affected by sequence dependent changeovers, this is a large number of integers. If there are 50 periods and 1000 SKUs that must be modeled as having sequence dependent changeovers, then this model will be way too large for standard optimization software to handle in a straightforward way. The dramatic increase in model size might not be worth the increase in accuracy. But in some situations, bottleneck resources have significant changeover time requirements and different sequences result in dramatically different capacity utilizations. The models are even more complex if it is necessary to also capture the ability to continue production of an SKU across a time bucket boundary. In such situations, it may be necessary to use heuristic search methods to address the sequencing problems (see §8.4).

4.4.5 A Few Remarks About Changeovers

Attempts at proper modeling of changeovers illustrate that modeling is largely an art form. At some point, we must make simplifying assumptions. There are more complications than the ones described here. The ones we have described should help readers to formulate constraints that capture the essence of their particular situations.

The process of developing the models can provide spinoff benefits. For example, we can gain a small but useful insight by considering the difficulties imposed by changeovers. It has always been obvious that an increase in the amount of time available for production of parts would result from engineering changes to production processes that reduce the time required for each changeover. Furthermore, the nuisance of changeovers is felt by production management as well as shop floor workers. Over the past few decades there has been a significant engineering effort worldwide associated with reducing the time required to perform changeovers. Many of the techniques are under the rubric SMED, which is an acronym for single minute exchange of die. As the name implies, many of the ideas in SMED involve redesign of jigs, fixtures, dies, software and other components of production equipment associated with changeovers.

There has been a significant decrease in the time required for changeovers on many machines due to SMED related efforts. This has resulted in the liberation of time previously spent on changeovers that can now be spent producing parts. This is important and valuable, but our models suggest that there are additional benefits. By eliminating or reducing changeover times using such techniques we not only gain greater capacity, we also create a production facility that is easier to manage.

This is good news not just for lazy managers. It is good news for ambitious managers as well. The elimination of changeover time as a significant factor in production planning makes the problem much easier for optimization software. In this context, "easier" means that optimization can be performed much, much faster. This, in turn, means that the optimization can be performed more often as new information becomes available and this results in plans that are more up-to-date and more responsive to current customer needs and shop-floor realities.

Our models clearly demonstrate the improved management characteristics that result from setup reductions. As the models become simpler, they are easier to solve. But these simplicity benefits are present whether formal mathematical models and computers are used to find good production plans or if the plans are hashed out on a chalkboard. Good production plans are simply easier to obtain when changeover times are reduced.

The changeover modeling process can be a bit complicated, but yields important benefits. We can summarize as follows:

- Changeover time reduction is good not only because of increased capacity but also improved manageability;
- meanwhile modeling enables leveraging computing power to obtain solutions as well as the ability to gain insights via the models and the modeling process.

We return to issues related to creating and using production planning models in a supply chain. In order to make the right changeover decisions we need some cost information and perhaps alternative routing information. This is discussed in the next chapters.

5. A Better Model

The data processing approach to MRP II is to use a straightforward algorithm to create an mrp solution and then to see if the plan is feasible with respect to capacity. In keeping with a data processing or information technology mentality, the users can also be given a lot of information to help them change the input data. The optimization approach that we are developing here asks for information about objectives and tries to obtain a good or optimal plan automatically.

5.1 A Cost Based Objective Function

In order to create a good or optimal decision, we need an objective. The first issue that must be resolved is a very important one: are we trying to minimize costs or maximize profits? For now, we are trying to develop an improved MRP II model, so it seems best to try to minimize costs. This is consistent with the mrp philosophy that demands are given to production planners and their job is to meet them. In general, in this book, we pursue models that minimize costs.

5.1.1 Costs

When constructing the objective function, we make use of *marginal costs.* That is, we look only at costs that will change as a result of the decisions under our control. Perhaps this is clarified by giving some examples of costs that we do not include. The cost of raw materials are not included because the expenditures must be made regardless of the plan chosen (one could quibble that the timing of the cash flows would be effected, but the inclusion of holding costs captures most of this and besides, it is small compared with the other issues.) In most situations, there is no reason to include regular-time labor for the same reasons. The personnel planning process is typically separate from the relatively short-term production planning processes that are the subject of our models.

A minimal set of cost data is described in Table 5.1. The cost of holding an inventory item for one period is a discount rate times the cost of the item.

$H(i)$	Per period holding cost for SKU i
$C(i)$	Total (out of pocket) changeover cost for SKU i

Table 5.1. Basic Cost Data for an Improved MRP II Objective Function

Since inventory is a somewhat risky asset the discount rate should be at least as high as the firms weighted average cost of capital and perhaps higher. For example, a value of 25% per year is not unreasonable for discounting many types of inventory. So if the time buckets are weeks and the annual discount rate is 26% and the item costs \$100, then $H(i)$ is

$$\frac{0.26/\text{year}}{52\ \text{buckets/year}} 100\$/\text{item},$$

which is 0.5 dollars per item per time bucket.

The out-of-pocket cost for a changeover is also fairly straightforward as soon as we get past the difference between a *setup* and a *changeover*. If a machine or workgroup must spend the same amount of preparation time per item of SKU i, then we call this a setup and we can add the capacity utilization into our calculations of U. Any material that is wasted for each batch should be rolled into R.

As a final means of clarification, we remind the reader of the notion of *sunk costs*. This refers to costs that will be incurred regardless of the decisions under consideration. For example, if a setup is required before production of an item that has no substitutes, then costs associated with that setup are sunk as soon as we accept the demand. (We are ignoring the timing of such cash flows.) Costs that are incurred when we switch over to production of the item, but will not be incurred again during the production run without regard to quantity are not sunk, since we can create a plan that combines orders for an item so as to save changeover costs.

We must also develop marginal cost data for changeovers. The idea of avoiding the inclusion of sunk costs applies here as well. For changeovers, however, the calculations are a bit more subtle.

When we compute $C(i)$, we add in the marginal direct labor costs on all resources that are changed over for SKU i and the cost of all material wasted and energy used as a result of the changeover. The productive capacity needed for wasted material is taken care of by the waste data element, W, but the cost of this material should be added to C. Should we add in the cost of the machine? Generally, no. This will raise the eyebrows of accountants, but with some patience they can be convinced.

What is the real difference in cost between solutions that differ by one changeover? This is an important question. Suppose that a very expensive machine that takes up a lot of floor space is hardly ever used; i.e., suppose it is used about one hour per week. What will be the difference in cash position if the machine is down for changeover for an extra two hours? Almost none. An extra three hours? Almost none. The cost of the machine and the corporate overhead rates are not important in making the marginal decision to perform a changeover. The variable cost of labor and perhaps energy or scrapped material are the only relevant costs.

Now suppose some machine is heavily used. What if we plan for it to be down during some bucket for an extra hour of changeovers? How much does this cost us? It depends. If we happen to be lucky and the resulting schedule has no extra inventory or extra tardiness, then again it costs us nothing beyond the out-of-pocket labor, energy, and material costs associated with a changeover. If it results in a plan with extra holding, then the cost should be picked up by the H term in the objective function. If it results in extra tardiness (a more likely case), then we should pick up the cost in the A term. One is tempted to charge an opportunity cost of using a scarce resource for changeovers. However, what is the lost opportunity? If there is one, it was the opportunity to make something in time to meet a customer demand in a timely fashion.

In an environment with increasing demands, it is reasonable to assess an extra penalty to tardiness in the last period to capture the notion of lost sales. The modeling for this is formalized in §6.3. The idea is that if there are significant changeover times required, but not enough capacity to meet demand, the capacity can be used either for changeovers or for production. Presumably if it is optimal to make more changeovers then it is optimal to avoid tardiness during the planning horizon. If the cost of lost sales is too high, then it is optimal to plan for fewer changeovers with an associated increase in productive capacity.

Of course, other arguments can be made and other cost structures can be used. In the end, it is up to the modeler to provide costs that result in good decisions. The idea behind using marginal costs is that they anticipate the difference between potential decisions.

5.1.2 Objective Function

Armed with these cost estimates, we can formulate a cost minimizing objective function. Before we do, we will find it useful to introduce the idea of *macros* or *notational variables*. These will not really be variables, they will be simple functions of variables that we will create to make it easier to read the models. We have used notation like $I(i, 0)$ for data and notation like $x_{i,t}$ for variables. So data have the list indexes in parenthesis and variables have them as subscripts. For macros, we will show the list indexes as subscripts and the variables on which the macro depends will be shown in parenthesis.

We need to emphasize that these macros behave like typical computer macros, which is to say that they rely on the computer to replace the macro literally with the corresponding string. They are not new variables or data. We show the macro definition using $\equiv$. The first macro we will use denotes the inventory of SKU i in period t:

$$I_{i,t}(x,\delta) \equiv \sum_{\tau=1}^{t-LT(i)} x_{i,\tau} + I(i,0) - \sum_{\tau=1}^{t}\left(D(i,\tau) + \sum_{j=1}^{P}(R(i,j)x_{j,\tau} + W(i,j)\delta_{j,\tau})\right)$$

We then write the demand and materials requirements constraint for all times t and all SKUs i as

$$I_{i,t}(x,\delta) \geq 0$$

with the understanding that this will be interpreted as

$$\sum_{\tau=1}^{t-LT(i)} x_{i,\tau} + I(i,0) - \sum_{\tau=1}^{t}\left[D(i,\tau) + \sum_{j=1}^{P}(R(i,j)x_{j,\tau} + W(i,j)\delta_{j,\tau})\right] \geq 0.$$

In other words, we use the macro $I_{i,t}(x,\delta)$ to be able to write exactly the same constraints that appear in model **MRPII** in a much more condensed form. Combining the macro expression with $I_{i,t}(x,\delta) \geq 0$ results in the same constraint.

Note that I always represents inventory, but $I(i,0)$ is given as data, while $I_{i,t}(x,\delta)$ is a macro. We use the macro to give us a shorthand to write it, but we should bear in mind that $I_{i,t}(x,\delta)$ represents a complicated expression for the planned inventory of SKU i at time t.

We can use the macro to write our objective to be the minimization of

$$\sum_{t=1}^{T}\sum_{i=1}^{P}[H(i)I_{i,t}(x,\delta) + C(i)\delta_{i,t}]$$

subject to the MRP II constraints. Inspection of this objective function indicates that the changeover cost, C, should include the cost of wasted material. The data element W allows us to take into account the wasted material that will have to be produced but does not "charge" us for the extra money spent; however, by adding the term $W(i,j)\delta_{j,\tau}$ to the inventory expression, we are "charged" for the capacity used.

The term $H(i)I_{i,t}(x,\delta)$ could have been written as

$$H(i)\left(I_{i,t}(x,\delta) + LT(i)x_{i,t}\right)$$

to add the cost of holding work in process inventory (WIP) to the objective function. The term $H(i)I_{i,t}$ includes only the cost of those SKUs that have been finished but not yet used or shipped to a customer. Since $H(i)$ gives the cost of holding one item in inventory for one time bucket, it can be multiplied by the term $LT(i)x_{i,t}$ to get the cost of holding the SKU during the lead time (i.e., while the SKU is being made). Before we quibble over whether there should be one cost for holding an SKU when it is finished and another for holding it while it is WIP, we should note that for this model the WIP holding costs are sunk. Once we have committed to the demand data, we have committed to all of the implied production. We have modeled the lead time as constant data, so this cost will not vary with the decision. Consequently, we do not include it at this point. In more sophisticated models, such as the one given in §6.1 where substitute components and multiple routings are considered, this term must be added to the model.

Now is a good time to reflect on the differences between the cost based objective functions that we have developed and the as-late-as-possible objective function implied by mrp and MRP II. The developers of mrp did not use an objective function or the terminology of optimization, but mrp and MRP II result in plans that correspond to the as-late-as-possible objective function. The philosophy behind this is motivated partly by a desire to keep holding costs at a minimum. The usual operational descriptions of mrp and MRP II obscure the simple objective function that is implied.

Our philosophy of explicitly writing down an objective function results in numerous advantages. Distinctions between the holding costs associated with each SKU can be noted. We are also able to consider other costs as well and could use a profit maximizing objective function when that is appropriate. These benefits accrue in addition to the important benefit of simultaneously considering the objective function while finding a capacity feasible plan.

5.2 Overtime and Extra Capacity

It is often overly simplistic to impose a single, hard capacity constraint for each resource in each period. This is because it is often the case that capacity can be added, particularly with some advance notice as would be the case if the production plan called for it. Classic examples are the use of overtime or an extra shift. In other cases, it may be possible to bring in extra resources on relatively short notice. To keep our language simple, we will refer to all such short term capacity additions as overtime. We now extend the MRP II constraints to capture this notion in a number of different ways.

5.2.1 A Simple Model

In order to produce a model that can be used easily in a variety of settings, we continue with the convention of working with capacity fractions rather than absolute measures of capacity. The data needed are described in Table 5.2. To understand these data, let us consider again the example developed in

§4.1. If indeed HR-101 is a resource of ten people who each work eight hours in each time bucket and if they can be asked to work up to four hours of overtime at a cost of \$10 per person extra per hour during the overtime, then we can calculate the needed data for HR-101. Remembering that HR-101 is resource number one and has eighty hours of capacity per time bucket, we write that $F(1,t) = 40/80 = 1/2$ and that $O(1,t) = \$10 \times 80$ for all time buckets t. The value of O is the extra cost doubling the capacity of the resource at overtime rates. This is the correct number to use even if doubling is not possible because it gives the right cost per fraction of capacity added.

$F(k,t)$	Maximum fraction of resource k that can be added in t
$O(k,t)$	Marginal cost per fraction of resource k added in t

Table 5.2. Data Short Term Capacity Expansion

We must introduce a new variable, $y_{k,t}$, to represent the overtime fraction for resource k in period t. This allows us to add the term

$$\sum_{k=1}^{K} O(k,t) y_{k,t}$$

to the objective function to capture the cost of overtime. We must add constraints to establish the overtime fraction for all resources k and times t and then constrain overtime and capacity:

$$\sum_{i=1}^{P} U(i,k) x_{i,t} \leq 1 + y_{k,t}$$
$$y_{k,t} \geq 0$$
$$y_{k,t} \leq F(k,t)$$

The capacity constraint is shown here without changeovers. They can be added if needed.

5.2.2 Complications

Yet again we have a need for some modeling art. It is never really the case that some arbitrary amount of overtime can be requested, but that is what this model implies. To avoid such issues, integer variables could be used to force the overtime plans to be for only certain numbers of hours. However, this is often not worth it. The additional integer variables slow down the solver for almost no managerial benefit. Remember that what will be used for planning is the production quantities and an “anticipated” need for overtime.

On the other hand, there may be cases when extra capacity can only be added in large amounts or with a very large fixed cost. When this is the situation, an integer variable must be added to indicate that overtime is used. This new variable will have exactly the same relationship to the y variables that production indicators, δ, have to the production quantity variables, x. It can then appear in the objective function to reflect the fixed cost or in a constraint to force the y variables to be above some value.

Further complications include the possibility that overtime has to be added for multiple resources at once. This might be the case when one crew operates a number of machines or when an audit team either works together or not at all. This is most directly handled by changing the subscripts on the y variables. Each resource group that must share overtime is grouped into a set. We will now introduce notation for these sets. Suppose there are E such resource group sets organized into a list of sets called $\mathcal{E}$. The variable list y will not be indexed by k going from one to K, but by e going from one to E. Instead of writing the capacity constraints as

$$\sum_{i=1}^{P} U(i,k)x_{i,t} \leq 1 + y_{k,t} \qquad k = 1,\dots,K, \ \ t = 1,\dots,T$$

we generalize them to be

$$\sum_{i=1}^{P} U(i,k)x_{i,t} \leq 1 + y_{e,t} \qquad e = 1,\dots,E, \ \ k \in \mathcal{E}_e, \ \ t = 1,\dots,T$$

Since this notation is unfamiliar, let us give a concrete example. Suppose that there are only two resources and they share overtime. Then E is one and $\mathcal{E}_1 = \{1,2\}$ so the constraints will be of the form:

$$\sum_{i=1}^{P} U(i,1)x_{i,t} \leq 1 + y_{1,t} \qquad t = 1,\dots,T$$

$$\sum_{i=1}^{P} U(i,2)x_{i,t} \leq 1 + y_{1,t} \qquad t = 1,\dots,T$$

To see why we say that this "generalizes" the original constraints, note that if $E = K$ and every $\mathcal{E}_e$ consists only of the index e, then the constraints will be the same.

5.3 Allowing Tardiness

For the purposes of mathematical modeling, it is useful to make a clear distinction between *deadlines* and *due dates.* A deadline is modeled using constraints, while a due date appears only in the objective function. The clear implication is that deadlines *must* be respected while due dates are more of a target.

The word "deadline" sounds a bit extreme, since it is extremely unusual for a customer to actually kill someone if their order is late (although not at all unusual for them to threaten it). For end items, deadlines are typically the correct model only for items that are of the highest priority. Other items can be allowed to be tardy and a penalty function can be used to model the fact that tardiness is more of a problem for some SKUs and/or for some customers.

Deadlines are the appropriate way to model intermediate SKUs. Since we are creating a plan, it makes no sense to plan for the production of an end item to begin before the completion of its components. Consequently, the completion times for items that will be used in assemblies must be enforced in the constraints in order for the model to make sense. This statement may not always be exactly what is needed (e.g., if there are end items that require some components initially and others a few time buckets later.) This is an advanced topic, but can typically be addressed by splitting the item into two SKUs one of which is made entirely using the other. Our concern now is to make a simple model that allows end items to be tardy but requires components to be on time.

5.3.1 A Simple Model

We introduce as data $A(i)$, which is the per period tardiness cost for external demand for SKU i. To make use of it, we need to make another significant change to the materials requirements and demand constraint. To allow tardiness, instead of using

$$I_{i,t}(x, \delta) \geq 0$$

we use

$$I_{i,t}(x, \delta) \geq -\sum_{\tau=1}^{t} D(i, \tau).$$

In other words, we allow a negative inventory position. Items with negative inventory positions are often referred to as *backordered*. The idea is that we still need to require production of items that are needed as components or else the entire plan is invalid. However, we want to allow the inventory position to be negative for an SKU to the extent that we have external demand for that SKU. This can be seen by expanding the macro $I_{i,t}(x, \delta)$ and then simplifying it to eliminate the summation over D. The result after expanding the macro and rearranging terms is:

$$\sum_{\tau=1}^{t-LT(i)} x_{i,\tau} + I(i,0) - \sum_{\tau=1}^{t} \sum_{j=1}^{P} (R(i,j)x_{j,\tau} + W(i,j)\delta_{j,\tau}) \geq 0$$

In order to construct an objective function that takes into account back-order quantities, we have to distinguish between negative and positive inventory positions. We let I^- be $-I$ if $I < 0$ and zero otherwise and let I^+ be I if $I > 0$ and zero otherwise. To get I^+ and I^-, add the following constraints for all i and t: $I^+_{i,t} - I^-_{i,t} = I_{i,t}(x, \delta)$, both $I^+_{i,t} \geq 0$, and $I^-_{i,t} \geq 0$. These constraints would allow I^+ and I^- to become arbitrarily large, but they will not do so because there will be positive costs associated with them in the objective function.

We can add terms like the following to the objective function:

$$\sum_{t=1}^{T} \sum_{i=1}^{P} (A(i)I^-_{i,t} + H(i)I^+_{i,t})$$

5.3.2 Complications

Many of our cost structures developed thus far depend on the SKU. For many SKUs and many customer orders, these costs are good enough to drive the planning process. However, there are always a few special orders, or special time periods. For example, it might be extremely important to make sure that an order from a new customer for item AJ8172 is delivered on time. Although the tardiness cost assigned to AJ8172 might generally be about right, it might be way too low for capturing the importance of this particular order. A large amount of future business might be lost if the order is late and this needs to be captured in the cost structures. It is important to distinguish not just among products, but among orders in this type of situation.

If it is necessary to differentiate among orders, "behind the scenes" SKUs are a direct way to accomplish this; for example the SKU AJ8172z could be used for a special customer or even a special customer order. For the case of special orders, A can be extended to include a period number so that $A(i,t)$ gives the tardiness cost in period t. This means that the objective function can be severely penalized for any solution where these orders are excessively late. Furthermore, special customer orders that will not cause problems if they are only one or two periods late can be given special SKUs and special low $A(i,t)$ values for t values not much beyond the due date. Adding SKUs can slow down the solution process, but if used in moderation, the solution quality can reflect important considerations and tradeoffs. If we carry the process to its extreme by considering every order to be deserving of its own SKU, then we have created a scheduling problem with tagged orders. A proper planning model should have enough scheduling detail so as to be realistic, but not so much that it cannot be run for a large enough portion of the supply chain to get the big picture. To say this in another way: the model will get to be too big unless we use this notion of an extra SKU for special customers only when it is very important to the planning – as opposed to the scheduling – process. A simple example of a scheduling subproblem embedded in a planning model is given in §8.4.3.

The trouble with using A values is that it is not clear how to assign the values. How much does it cost to be tardy? This difficulty and other objective function issues are discussed in the next section. A more drastic measure is to make A a non-linear function, but this is not consistent with our desire to use linear models. Non-linear models are discussed briefly in later chapters.

5.4 Objective Function Issues

If we ignore the discussions that we have labeled as complications, we have settled so far on an objective of minimizing marginal costs using a function of the form

$$\sum_{t=1}^{T}\left[\sum_{i=1}^{P}\left(A(i)I_{i,t}^{-}+H(i)I_{i,t}^{+}+C(i)\delta_{i,t}\right)+\sum_{k=1}^{K}O(k,t)y_{k,t}\right].$$

We discussed the holding cost, H, and the changeover cost, C, in §5.1.1. Of the two new cost elements, A is the toughest data element to estimate. The value for O is not quite so difficult, so that is where we begin our discussion.

The data element O gives the marginal cost of adding a unit of extra capacity. Its calculation is complicated only by the use of fractions as a measure of capacity. The easiest way to compute a value for $O(k,t)$ is to pretend that it is possible to double the capacity of resource k and use the marginal cost of doing so (for all time buckets t).

For example, if a resource is usually available for eight hours per day at a cost of \$25/hour and overtime is available at a cost of \$37.50/hour, then $O(k,t)$ would be $(8)(12.50) = 100$. This means that if $y(k,t)$ is 0.25 for some time t (which would imply 2 hours of extra capacity) then the marginal cost would be \$25.00. We are back to the issues associated with sunk and marginal costs that we first broached in §5.1.1.

Notice that we do not compute a cost of using resources unless they are used in "overtime." As the model is presently constructed, this makes sense because the use of the resources is a sunk cost; that is to say, the production must occur so the costs will be incurred no matter what decision we make (the time value of money for resource allocations is essentially captured in the holding costs.) When we consider the possibility of alternate routings in §6.1, we will have to take into account the potential for costs that differ according to the routing.

This brings up an important point. If the costs vary dramatically from one period to the next, then it would be necessary to include production costs in the objective function. In general we try to avoid putting sunk costs in the model, but this means that we should take care to include all marginal costs that vary with the decision variable values.

Setting a value for $A(i)$ is not straightforward. The value for A probably should be at least the gross margin for SKU i based on an opportunity cost argument. When an item of SKU i is late, an opportunity is missed to supply a customer who wanted the item in a time bucket. The capacity that would have been used for that product is presumably used for changeovers or some other product. It is a bit overly simplistic to argue that we have created a vacuum in the market for SKU i that will be filled by a competitor (perhaps in some unseen way), but only a bit. Tardiness, and a customer's response to it, is the invisible hand's way of correcting overutilization of capacity. For strategic reasons, an even higher value may be needed to reflect loss of goodwill or loss of market share. If very short time buckets are used, it may be unrealistic to assess a high penalty for being only one period late. In this case it might be necessary to use some form of non-linear penalty as described in §8.3.

There will be more discussion of the objective function later, but before leaving this topic we note that using a profit maximizing objective function will require some changes to the constraints as well as the objective function. The changes to the model are not major. The cost terms can remain, but must be subtracted from a term that gives the profit associated with each SKU. In a profit maximizing environment, presumably not all demands are firm. That is, some of the demand constraints should be given as $\leq$ rather than $\geq$. The advantage of a profit maximizing model is that the product mix can be selected automatically. We can invert this argument. If the product mix is not a production decision, then a profit maximizing objective function probably is not sensible.

5.5 The Model

We now give the base model as developed in this chapter. In actual practice, many of the data elements will be zero and some of the constraints and variables can be omitted altogether. The complete picture is given here to summarize the chapter in one concise statement. We refer to the model as **SCPc** to emphasize that it is a cost minimization model for supply chain production planning. The data are given in Table 5.3 and the model is shown as Figure 5.1 making use of the macro shown in Figure 5.2. The variables that are to be set by the model are shown in Table 5.4. In some cases, the constraints have been rearranged to put a zero on the RHS. The modifications listed as complications are not shown here.

The use of the symbol I requires a little explaining. We use the symbol consistently to represent inventory position, but that requires it to appear with the data, variables and the macros. The initial inventory, $I(i, 0)$, is given as data. The inventory expression depends on existing variables and data so it is given as a macro. Backorder and positive inventory quantities have to be constrained, so they must be given as variables.

P	Number of SKUs
T	Number of time buckets
K	Number of resources
$LT(i)$	Lead time for SKU i
$I(i,0)$	Beginning inventory of SKU i
$D(i,t)$	External demand for SKU i in period t
$R(i,j)$	Number of i's needed to make one j
$U(i,k)$	Fraction of resource k needed by one unit of SKU i
$F(k,t)$	Maximum fraction of resource k that can be added in t
$S(i,k)$	Fraction of resource k used to changeover to SKU i
$W(i,j)$	Waste of SKU i to changeover to SKU j
$H(i)$	Per period holding cost for SKU i
$C(i)$	Total (out of pocket) changeover cost for SKU i
$O(k,t)$	Marginal cost per capacity fraction added to resource k
$A(i)$	Per period tardiness cost for external demand for SKU i
M	A large number

Table 5.3. Data for the **SCPc** Model

$x_{i,t}$	Order release quantity for SKU i in time t
$y_{k,t}$	"Overtime" fraction of resource k in time t
$\delta_{i,t}$	Binary indicator of production of SKU i in time t
$I^+_{i,t}$	Inventory of SKU i to carry in time t
$I^-_{i,t}$	Quantity of SKU i backordered in time t

Table 5.4. Variables Set by the **SCPc** Model

Minimize:

$$\sum_{t=1}^{T}\left[\sum_{i=1}^{P}\left(A(i)I^-_{i,t} + H(i)I^+_{i,t} + C(i)\delta_{i,t}\right) + \sum_{k=1}^{K} O(k,t)y_{k,t}\right] \qquad \textbf{(SCPc)}$$

subject to:

$$\begin{array}{lll}
I_{i,t}(x,\delta) + \sum_{\tau=1}^{t} D(i,\tau) & \geq 0 & i=1,\ldots,P,\ \ t=1,\ldots,T \\
\sum_{i=1}^{P}\left[U(i,k)x_{i,t} + S(i,k)\delta_{i,t}\right] & \leq 1 + y_{k,t} & t=1,\ldots,T,\ \ k=1,\ldots,K \\
y_{k,t} & \leq F(k,t) & t=1,\ldots,T,\ \ k=1,\ldots,K \\
\delta_{i,t} & \geq \frac{x_{i,t}}{M} & i=1,\ldots,P,\ \ t=1,\ldots,T \\
\delta_{i,t} & \in \{0,1\} & i=1,\ldots,P,\ \ t=1,\ldots,T \\
x_{i,t} & \geq 0 & i=1,\ldots,P,\ \ t=1,\ldots,T \\
y_{k,t} & \geq 0 & t=1,\ldots,T,\ \ k=1,\ldots,K \\
I^+_{i,t} & \geq 0 & i=1,\ldots,P,\ \ t=1,\ldots,T \\
I^-_{i,t} & \geq 0 & i=1,\ldots,P,\ \ t=1,\ldots,T \\
I^+_{i,t} - I^-_{i,t} & = I_{i,t}(x,\delta) & i=1,\ldots,P,\ \ t=1,\ldots,T
\end{array}$$

Fig. 5.1. SCPc Model

$$I_{i,t}(x,\delta) \equiv \sum_{\tau=1}^{t-LT(i)} x_{i,\tau} + I(i,0) - \sum_{\tau=1}^{t}\left(D(i,\tau) + \sum_{j=1}^{P}\left(R(i,j)x_{j,\tau} + W(i,j)\delta_{j,\tau}\right)\right)$$

Fig. 5.2. SCPc Inventory Macro

6. Extensions to the Model

In the previous chapter we extended well beyond the data processing concepts embodied by MRP II to make use of the capabilities of an optimization model. In this chapter, we continue this course by describing a number of important extensions to the model. Depending on the planning environment, some or all of these extensions may be needed to produce a useful model.

We consider first an extension that captures the idea that SKUs can be routed through a variety of production facilities or subcontractors. We follow this with extensions to the model that penalize changes to the plan so that the planned production from one "optimal" plan to the next does not vary wildly. A related topic is the handling of end-of-horizon effects. That is, we may need to modify the model so that it does not result in a plan with strange production quantities and inventory positions in the last period. Finally, we provide some extensions to support modeling of transportation and product movement.

6.1 Substitutes, Multiple Routings and Subcontractors

Tactical production planning in a supply chain involves more than just determining production quantities or authorizing overtime. Often, choices must be made between alternative modes of production when there are multiple routings, multiple vendors, or subcontractors for parts that are also produced in-house. Each of these may have a different cost and lead time.

There are choices to be made between suppliers, or between different factories or between different production lines in the same factory. In this situation, we would say that multiple routings are available. We would like to create a model that picks between the alternatives in an optimal way while simultaneously considering the effect on all other decisions regarding production and capacity utilization.

Furthermore, a single supplier might have multiple offerings that can be used to create the needed SKU. These multiple offerings might differ in minor physical ways or they might be exactly the same part but with different lead times and different costs. For example, sometimes a larger size can be substituted for a smaller one with no loss in functionality or two SKUs can serve the same purpose, but one can make use of material of a higher quality than necessary. In other cases, suppliers offering expedited delivery can result in a shorter lead time at higher price. If we think of the lead time as one of the important attributes of the SKU, then it is natural to consider an SKU shipped sooner to be a substitute for the SKU with the standard lead time. In general, a different routing or the use of a substitute part will involve a different lead time and this can be important to model.

Another possibility is that the SKU can be made either "in-house" or by an outside subcontractor. Generally, both the cost and the lead time will be different for subcontractors. With just a bit of abstraction, we can see that the choice between subcontractors or the decision regarding whether to use a subcontractor is the same as choosing between alternative routings or selecting one SKU from among a list of substitutes.

With some effort, all three of these possibilities can be added to our models using the same basic mechanisms. Matters are complicated somewhat by the need to allow for differing lead times while at the same time allowing parts with substitutes or alternative routings to appear anywhere in the bill of materials. In order to do this, we propose changes to the model as well as additional data structures.

One possibility to model such situations is to have a separate SKU for each substitute, alternate routing and/or for each subcontractor. So if AJ8172 could go along two routings, there would be SKUs AJ8172a and AJ8172b. This results in the need for a new data structure, the list of substitutes sets, $\mathcal{L}$. This list has P elements in it. Each element of $\mathcal{L}$, $\mathcal{L}(i)$, is a set of substitute SKU indexes for the SKU i. For those SKUs that do not have substitutes the substitutes list is empty, which we write as $\mathcal{L}(i) = \emptyset$.

We assume that regardless of the substitute used, the substitute parts are essentially indistinguishable upon completion for the purpose of computing holding and tardiness costs. This means that they can be accounted for after completion using a single SKU that we will refer to as *master* SKU. To use mathematical notation, master SKUs are those SKUs i for which $\mathcal{L}(i) \neq \emptyset$. To make it convenient to distinguish SKUs that are in an alternates list for at least one master SKU, we use the unusual notation $j \in \mathcal{L}$ to signify that SKU j is in the list $\mathcal{L}(i)$ for some master SKU i.

We now continue with the example where SKU AJ8172 has substitutes AJ8172a and AJ8172b. Suppose as before that AJ8172 is SKU 1. Further suppose that the substitutes are SKUs 6 and 7, respectively. Under these conditions $\mathcal{L}(1) = \{6, 7\}$ and both $j = 6$ and $j = 7$ would be j values with the property $j \in \mathcal{L}$.

Note that the value of P includes the substitute SKUs. Our expressions are cleaner if we simply treat the master SKUs and substitutes the same as all other SKUs. We rely on data such as the substitutes list to provide the distinctions.

Master SKUs are constrained to have zero production (i.e., for master SKU i, $x_{i,t} = 0$ for all t). But we insist that all requirements be for the master SKU. In other words, all external demands will be for the master SKU and any product that can use one of the substitutes as component will have an $R(i, j)$ value for the master only. This means that the inventory macro for master SKUs must be changed to

$$I_{i,t}(x, \delta) \equiv \sum_{\ell \in \mathcal{L}(i)} \sum_{\tau=1}^{t-LT(\ell)} x_{\ell,\tau} + I(i, 0) - \sum_{\tau=1}^{t} \left(D(i, \tau) + \sum_{j=1}^{P} R(i, j) x_{j,\tau} \right) .$$

So the production is done for one or more of the substitutes to meet the demand and requirements for the master.

We must pause to mention one complication that arises when implementing this on a computer. Simply changing the macro for master SKUs has notational advantages in the book, but does imply some effort when implementing. As a practical matter with most modeling languages, every constraint with the $I_{i,t}$ macro must be written twice: once for the master SKUs and once for the others. Furthermore, the meaning of "for the others" is not completely simple. The mrp constraint does not really make much sense for those SKUs that are substitutes because they cannot have external demand and cannot have non-zero requirements for use in other SKUs (all requirements for demands are for the master SKU). Consequently, as a practical matter one does not want to have a mrp constraint for substitutes. It is not conceptually difficult to implement this model, but having a different $I_{i,t}$ macro for different types of SKUs does require some attention.

Although many organizations do not have material waste associated with changeovers and consequently would not use W values at all, in the interest of being thorough we must concern ourselves with how these data are treated in a model that uses substitutes and master SKUs. Since there is no production of master SKUs, the term $W(i, j)\delta_{j,\tau}$ is not needed in the macro for a master SKU. Furthermore, $W(j, j)$ is always zero for master SKUs j because master SKUs are not produced. However, $W(i, j)$ could be greater than zero for some master SKU i and some j if the SKU happens to be wasted during a changeover to SKU j. Also note that for an SKU i that is a substitute, if $W(i, i)$ is non-zero then one needs to subtract $W(i, i)\delta(i, t)$ from the x value in the production summation term for the corresponding master SKU macro because otherwise the waste of the substitute would not be accounted for.

The original inventory macro can be used for the SKUs for the substitutes themselves as well as for SKUs that do not have substitutes. However, the inventory position variables for the substitutes, I^+ and I^-, must be fixed at zero. This will enforce the idea that inventory is held only as the master SKU. In the unusual situation where substitutes have significantly different holding costs, then additional binary variables must be used to indicate which of the substitutes is in inventory. This will not be developed here.

If costs differ for the substitutes, then terms must be added to the objective function. Once again, we are reminded that we are basing our models on marginal costs so the lowest cost substitute will have a cost of zero and the others will have cost that is the difference between their cost and the lowest cost substitute. The idea is that as soon as we accept the order a cost at least as high as the lowest cost substitute is sunk. The marginal costs of substitutes should include any significant difference that arise as a result of holding cost differences that result from widely varying lead times.

The additional data required are shown in Table 6.1. The full formulation for alternative routings as an extension to problem **SCPc** is given as **SCPa** in Figure 6.1 where the shorthand notation $\forall$ and $\in$ is abused in the interest of brevity to mean that we want indexes for all appropriate values of the indexes listed. For example, $t, \ell \notin \bigcup_{i=1}^{P} \mathcal{L}(i)$ means the constraints are repeated for $t = 1, \ldots, T$ and for values of ℓ in the set formed by removing from $1, \ldots, P$ the indexes that appear somewhere in the entire list of substitute sets $\mathcal{L}$.

$\mathcal{L}(i)$	Alternates (or substitutes) list for SKU i
$V(i)$	Marginal cost of using alternate SKU i

Table 6.1. Additional Data Required for Alternate Routings

The term

$$H(i)\left(I^{+}_{i,t}(x,\delta) + LT(i)x_{i,t}\right)$$

adds the cost of holding finished inventory as well as work in process inventory to the objective function. If the cost of WIP for an SKU is significantly different than the cost of completed items, then a different cost element must be added, $H'(i)$ and the term is split into:

$$H(i)I^{+}_{i,t}(x,\delta) + H'(i)LT(i)x_{i,t}$$

As was discussed on page 48, for the production system models we have considered so far, the cost of holding WIP will not vary with the decisions. However, we include it here because it will be needed in more sophisticated models that build on this model.

6.2 Penalizing Changes to the Plan

In many production environments, a stable plan is almost as important as a good plan. It is bad if a new plan is generated weekly and the planned production for a given week changes radically each time a new plan is generated. This causes production workers to wonder if there is something wrong with the planning process.

Minimize:

$$\sum_{t=1}^{T}\left[\sum_{i=1}^{P}\left(A(i)I_{i,t}^{-}+H(i)\left(I_{i,t}^{+}+LT(i)x_{i,t}\right)+C(i)\delta_{i,t}+V(i)x_{i,t}\right)\right.$$
$$\left.+\sum_{k=1}^{K}O(k,t)y_{k,t}\right] \qquad \textbf{(SCPa)}$$

subject to:

$$\begin{array}{lll}
I_{i,t}(x,\delta)+\sum_{\tau=1}^{t}D(i,\tau) & \geq 0 & i=1,\ldots,P,\ t=1,\ldots,T\\
\sum_{i=1}^{P}\left[U(i,k)x_{i,t}+S(i,k)\delta_{i,t}\right] & \leq 1+y_{k,t} & t=1,\ldots,T,\ k=1,\ldots,K\\
y_{k,t} & \leq F(k,t) & t=1,\ldots,T,\ k=1,\ldots,K\\
\delta_{i,t} & \geq \frac{x_{i,t}}{M} & i=1,\ldots,P,\ t=1,\ldots,T\\
\delta_{i,t} & \in\{0,1\} & i=1,\ldots,P,\ t=1,\ldots,T\\
y_{k,t} & \geq 0 & t=1,\ldots,T,\ k=1,\ldots,K\\
x_{\ell,t} & \geq 0 & \forall\, t,\ell\in\bigcup_{i=1}^{P}\mathcal{L}(i)\\
x_{\ell,t} & = 0 & \forall\, t,\ell\notin\bigcup_{i=1}^{P}\mathcal{L}(i)\\
I_{\ell,t}^{+} & = 0 & \forall\, t,\ell\in\bigcup_{i=1}^{P}\mathcal{L}(i)\\
I_{\ell,t}^{-} & = 0 & \forall\, t,\ell\in\bigcup_{i=1}^{P}\mathcal{L}(i)\\
I_{\ell,t}^{+} & \geq 0 & \forall\, t,\ell\notin\bigcup_{i=1}^{P}\mathcal{L}(i)\\
I_{\ell,t}^{-} & \geq 0 & \forall\, t,\ell\notin\bigcup_{i=1}^{P}\mathcal{L}(i)\\
I_{i,t}^{+}-I_{i,t}^{-} & = I_{i,t}(x,\delta) & \forall\, t,\ell\notin\bigcup_{i=1}^{P}\mathcal{L}(i)
\end{array}$$

Fig. 6.1. SCPa Model

Such concerns may be well-founded. If the optimal plan produced one week differs greatly from the plan produced the preceding week, production workers will learn to ignore the plan and make use only of the short term schedule. When this happens, the value of planning will be significantly reduced. We can think of a reduction in the value of planning as a cost and capture the cost directly in our model by penalizing changes to the plan.

The model is extended by taking the values of the variables from the last plan generated as data. Terms are added to the objective function that penalize the absolute value of the difference between the value that the variable will take in the current plan. We use $X(i,t)$ as data to indicate the values that were assigned to the production variables during the last planning phase and $E(i,t)$ to give the penalty for deviating from the last plan during the next planning cycle. In order to penalize both negative and positive deviations, we create the variables X^{-} and X^{+} that play a role similar to I^{-} and I^{+}.

We let X^- be $X - x$ if $x < X$ and zero otherwise and let X^+ be $x - X$ if $x > X$ and zero otherwise. To get X^+ and X^-, add the following constraints for all i and t:

$$X^+_{i,t} - X^-_{i,t} = x_{i,t} - X(i,t),$$
$$X^+_{i,t} \geq 0, \text{ and}$$
$$X^-_{i,t} \geq 0$$

These constraints would allow X^+ and X^- to become arbitrarily large, but they will not do so because there will be positive costs associated with them in the objective function. The objective function is modified to have the following additional terms:

$$\sum_{t=1}^{T}\sum_{i=1}^{P} E(i,t)\left(X^+_{i,t} + X^-_{i,t}\right)$$

Changes to the plan can be penalized in a fashion that is simpler and somewhat more realistic by using quadratic penalty terms, but this requires different software to solve the problem because the objective function will no longer be linear. Instead of making use of the variables X^- and X^+, the objective function would simply have the following quadratic terms added:

$$\sum_{t=1}^{T}\sum_{i=1}^{P} E(i,t)\left(x_{i,t} - X(i,t)\right)^2$$

Both problems are easier to solve if one penalizes changes only in real valued variables.

6.3 End-of-horizon Effects and Minimum Inventories

In general, and particularly when tardiness is allowed, some care must be taken to make sure that *end-of-horizon effects* are considered. With some cost structures, a computer that finds the optimal solution to the model might make some seemingly strange choices. For example, it might be optimal, for the given model, to simply not produce some items and to incur the tardiness costs every period. This is not a sensible plan, but it might be optimal for a model that must, of necessity, contain some simplifications.

A related difficulty is that the optimal solution is sure to have very low or zero inventories in the last period. The models we have developed so far provide no reason to produce goods that will still be in inventory after the last period. The only exception is that some inventory might remain when minimum lot sizes are being used. If it is not reasonable to produce plans with zero inventories, then the models must be altered to disallow them.

One technique is to add constraints that enforce end-of-horizon conditions. For example, the constraint

$$\sum_{\tau=1}^{T-LT(i)} x_{i,\tau} + I(i,0) - \sum_{\tau=1}^{T} \left(D(i,\tau) + \sum_{j=1}^{P} R(i,j) x_{j,\tau} \right) \geq 0$$
$$i = 1, \ldots, P$$

ensures that all demand must be satisfied by the end of the planning horizon. This line appears in the original mrp formulation, except there it is for all times, $t = 1, \ldots, T$, and here it is only for the final time period. In the mrp formulation, all demand had to be met at all times so such an end-of-horizon constraint would be redundant.

It can make more sense to have a big tardiness penalty for the last period, $A(i,T)$, instead of requiring all production to be finished by the end of the planning horizon. As mentioned in §5.1.1, this can be used to roughly approximate the cost of lost sales. This is particularly reasonable in an environment where demand is increasing. The demand that has not been satisfied by the end of the planning horizon can be thought of as demand that will not be satisfied because even more demand will be coming thereafter. The use of a high tardiness penalty for the last period is designed to force choices between using capacity for changeovers or for production.

Ending inventory should be reasonable, but for long time horizons it is often not worth it to require that a planner provides desired ending inventory positions for every item as input data. One way to avoid requiring excessive data requirements is to ask for a single fraction (e.g., 0.25) that will be used to restrict the range for the final inventory position. The constraints

$$(1 - IT)I(i,0) \leq I_{i,T}(x,\delta) \leq (1 + IT)I(i,0) \qquad i = 1, \ldots, P$$

restrict the final inventory to be within IT of the initial position for each SKU, where IT (for Inventory Tolerance) is a fraction that is provided as data.

From a modeling perspective, a simpler solution is for planners or managers to provide data, $I(i, T+1)$, specifying minimum ending inventory positions for each SKU i. One can add constraints such as

$$I_{i,T}(x,\delta) \geq I(i, T+1) \qquad i = 1, \ldots, P$$

to ensure that ending inventory positions are planned to be reasonable. End-of-horizon effects naturally become less important if the planning horizon T is longer than if it is shorter. As the planning horizon becomes longer, the importance of extremely good values for $I(i, T+1)$ is diminished because the effect on near term decisions is not as great. When minimum inventory levels have been established, these values are reasonable targets for the ending inventory.

Minimum inventory levels are often specified as a method of providing *safety stock*. Fluctuations in demand or in yields can result in shortages. To guard against these shortages, many manufacturers plan to maintain minimum inventory levels. Typically, these levels are set as a matter of policy and remain fixed over the planning horizon. When this is the case, we can use the data element $I(i, T+1)$ again to represent the minimum (if not, we must introduce a new data element). We then modify the constraint

$$I_{i,t}(x,\delta) + \sum_{\tau=1}^{t} D(i,\tau) \geq 0 \qquad i = 1,\ldots,P;\ \ t = 1,\ldots,T$$

to be

$$I_{i,t}(x,\delta) + \sum_{\tau=1}^{t} D(i,\tau) \geq I(i, T+1) \qquad i = 1,\ldots,P;\ \ t = 1,\ldots,T$$

It is reasonable to allow for backorders and at the same time plan for safety stock because the safety stock is an operational goal or target rather than a planning constraint. For end items, minimum inventory positions can be calculated using standard minimum service level formulas given in standard textbooks on production planning.

For the computer used to optimize the model, it is just a "game" to see how to find values for the variables that make the objective function as low as possible without violating any constraints. With some cost data (for example if work in process inventory costs are not present), production during the last periods can lower the objective function even though it might be the case that the lead time for the SKU would prevent this work from being completed before the end of the planning horizon. To avoid such a plan, we can add the constraints

$$\sum_{\tau=T-LT(i)+1}^{T} x_{i,\tau} = 0$$

for all SKUs i.

6.4 Modeling Product Movement and Transport

One important way to model transportation choices is using the modeling methods that were discussed in §6.1 on alternative routings. When an SKU can be made in a variety of locations, the marginal costs including the marginal cost of transportation, should appear in the objective function as multipliers for the production quantities. We now turn our attention to issues related to shipping along a fixed routing.

Like all modeling, transportation modeling can be simple or very complicated. Multiple locations can add a lot of computational complexity. The goal of the modeler is to capture the important features. In this section we explore a few extensions to our model to capture many essential features of product movement.

6.4.1 Simple Product Movement and Shipping

In the simplest case, there is a per unit charge for shipping or product movement that depends only on the SKU number and perhaps the time period. However, for our models this case is not interesting unless there are substitutes (see §6.1). Every SKU follows a fixed routing and any transportation costs will have to be incurred in order to satisfy demand so they are a sunk cost. If there are substitutes, then the differential cost of transportation should be combined with other differential costs and applied to the substitutes.

6.4.2 Expedited Shipping

An easy case, which happens to be common, is when the only important effect of production plans on transportation is the use of expedited shipping for end-items that would otherwise be late. This can be modeled in a reasonable way by simply making $A(i)$ the cost of expedited shipping. In other words, if shipping normally costs one dollar per piece that would not be included because it is a sunk cost as soon as we accept the order. However, if the expedited shipping costs ten dollars we could set $A(i)$ to be 9 for this item so the cost of being late would be captured. The model can be made more realistic (and much more complicated) by adding an indicator variable so that a much higher value for $A(i)$ would be used if the items would be so late that expedited shipping would not help.

The presence of multiple transportation options makes the modeling more difficult. For each transportation option, a resource is created. For each SKU that can use each transportation option an alternative routing is created as described in §6.1 using an alternate SKU. Each alternate SKU can then have the appropriate lead time for the mode of transportation. In order to put the marginal costs in the objective function, we assign a cost of zero to the cheapest option (and thereby omit it from the objective function). Other options are then included in the objective function using the difference in costs with respect to the cheapest alternative.

6.4.3 Fixed Costs and Consolidations

The main complications arise from fixed costs and consolidations. We use the words "fixed costs" here to refer to the situation where there is a fixed cost of transport between resources that must be paid for any amount shipped up to some maximum. This is the case when the product must be moved by a truck or must be placed in a container to be shipped or moved. Further complications result when multiple SKUs can be shipped together. This is often referred to as *consolidation*. There are many ways to model this, and some of them are beyond the scope of linear models. However, we sketch one approach here as an example.

We treat the means of transportation as a special resource and introduce the integer variables θ to capture the idea that we can use these resources in multiple, discrete units. For example, we can fill one, two, three, or more containers or we can use some number of trucks. For each SKU and each transportation resource the data $U(i,k)$ play the same role as always. They indicate what fraction of resource k is consumed by one unit of SKU i. We modify the capacity constraint for all transportation resources, k, to be

$$\sum_{i=1}^{P} U(i,k)x_{i,t} \leq \sum_{m=1}^{M_k} \theta_{k,m,t} \qquad t = 1, \ldots, T$$

where M_k is the maximum number of trucks or containers. It could be confused with the M used as a large number, but hopefully this is mitigated by use of the subscript k. The variables $\theta_{k,m,t}$ are intended to be assigned a value 1 in order so that the highest value of m for which $\theta_{k,m,t}$ is one indicates how many of the transport units of resource k are to be used in time bucket t. To enforce this, we add a series of constraints of the form

$$\theta_{k,m,t} \geq \theta_{k,m+1,t}$$

for all times t, all containers m from 1 to $M_k - 1$, and all transportation resources k. Finally, we add the constraints

$$\theta_{k,m,t} \in \{0, 1\}$$

for all transportation resources k, all times t, and $m = 1, \ldots, M_k$. There are other ways to handle this situation, but this method allows for graceful treatment of volume discounts. In anticipation of discounts, we also index the cost for each container of type k by m. We use $B(k,m)$ to be the cost of container m of type k. We then add the following terms to the objective function:

$$\sum_{t=1}^{T} \sum_{m=1}^{M_k} B(k,m)\theta_{k,m,t}$$

for all transportation resources k.

When there are no discounts, we could write $B(k,1) = B(k,2) = \cdots = B(k,m)$, but it is simpler to use an integer variable $\theta_{k,t}$ to indicate the number of containers used. So if there are no discounts, then θ does not need to be a list of binary variables indexed by m, but can simply be the number of containers. In this case, if the maximum number of containers used is large, then the model might be accurate enough if $\theta_{k,t}$ is allowed to take on fractional values (i.e., not required to be an integer so that $\theta_{k,t} = 77.2$ would be interpreted to mean "we plan for about 77 containers"). Whether or not $\theta_{k,t}$ is constrained to be an integer will have a big effect on the computational effort required to find an optimal solution, but either way the transportation term in the objective function is simplified to

$$\sum_{k=1}^{K} \sum_{t=1}^{T} B(k)\theta_{k,t}$$

and the transportation capacity constraint becomes

$$\sum_{i=1}^{P} U(i,k)x_{i,t} \leq \theta_{k,t}$$

for all times, t.

However, volume discounts for transportation are so common that generally if transportation is important enough to be modeled during planning, the full version with $\theta_{k,m,t} \in \{0,1\}$ is often needed.

6.4.4 Transportation Discounts

There are many types of transportation discounts, so we cannot hope to provide a model for all of them. We describe perhaps the two most common: marginal and all container discounts. If the discount structure is important and complicated, then perhaps linear models are no longer appropriate and heuristic search methods should be used (see §8.4).

In some situations, the cost for the second container or truck is lower than for the first and the third is lower yet. We refer to this as marginal container discounts. It can be captured simply by the correct values for $B(k,m)$ for each successive value of m.

All container discounts occur when the cost of shipping all containers depends on how many containers are shipped. In this case, we have to do a little arithmetic to calculate the marginal cost of each container. If we are going to ship some number of containers, m, then we are quoted a per container cost of shipping containers that we will call $B'(k,m)$. The total cost will be $mB'(k,m)$, but our objective function assigns a separate cost to each container, so we need to compute the marginal cost of each container. The correct value to use is computed as follows:

$$B(k,m) = mB'(k,m) - (m-1)B'(k,m-1)$$

This can be proven, but it may suffice for the reader to verify that the expression is correct for the first few values of k.

6.4.5 Discussion of Transportation Modeling

Optimized transportation planning and scheduling is a big topic. Under the rubric *logistics* a tremendous number of models and solution procedures have been created for all aspects of this problem domain. We could not hope to cover even a meaningful fraction of this topic, which warrants its own book. What we have done is to connect our earlier models with models that take transportation costs into consideration.

Some of the modeling of transportation is implicitly included in the work on alternative routings and substitutes (see §6.1). At the level of production planning in a supply chain, alternative routings is one of the most important decisions. When combined with an ability to model container discounts, the model can have enough detail for planning.

Since our intention is to develop a planning rather than a scheduling model, details like the time required to accumulate a container load (which might span time buckets) is probably too low level to include. There are, of course, potentially a host of details associated with transportation planning. Once again, we remind the reader that modeling is an art and the appropriate level of detail must be sought using a combination of insight into the operation of the business as well as experience gained while building up a model. This experience is best gained by beginning with a simple model and adding complications as deemed necessary.

We have also not modeled strategic issues such as facility location. The models that we have developed could be thought of as sub-problems for facilities location models. A model for strategic facilities location planning needs to be able to make use of a more detailed model to compute the cost associated with operating under a variety of location plans. The models that we have developed could serve to compute such costs. In other words, strategic planning models require tactical models to provide data that estimate the cost of operating under a particular strategy and set of locations. A tentative solution to the strategic planning model implies data for one of our models and the cost of a solution to the model provides data to the strategic planning model.

6.5 Summarizing the Model

We now provide a fairly complete model that contains most of the model features that we have discussed so far. In most situations, this entire model is not needed. It is listed here as a means of summarizing the models that we have developed so far. As such it is intended to serve as a reference.

To make the model pieces fit together, we will assume that the transportation resources have indexes above K and collect them in the set $\mathcal{K}$ as indicated in the consolidated list of data for this consolidated model given in

Table 6.2. The variables are listed in Table 6.3 and the inventory macros are in Figure 6.4 and Figure 6.3. The problem **SCPf** itself is shown in Figure 6.2.

P	Number of SKUs
T	Number of time buckets
K	Number of resources not for transportation
$LT(i)$	Lead time for SKU i
$I(i,0)$	Beginning inventory of SKU i
$I(i,T+1)$	Target ending inventory of SKU i
$D(i,t)$	External demand for SKU i in period t
$R(i,j)$	Number of i's needed to make one j
$U(i,k)$	Fraction of resource k needed by one unit of SKU i
$F(k,t)$	Maximum fraction of resource k that can be added in t
$S(i,k)$	Fraction of resource k used to changeover to SKU i
$W(i,j)$	Waste of SKU i to changeover to SKU j
$H(i)$	Per period holding cost for SKU i
$C(i)$	Total (out of pocket) changeover cost for SKU i
$O(k,t)$	Marginal cost per fraction added
$A(i)$	Per period tardiness cost for external demand for SKU i
$\mathcal{L}(i)$	Alternates (or substitutes) SKU list for SKU i
$V(i)$	Marginal cost of using alternate SKU i
$X(i,t)$	Production quantity of SKU i for period t for the last plan
$E(i,t)$	The penalty for deviating from the last plan
$B(k,m)$	The cost of container m of type k
$\mathcal{K}$	Set of transportation resource indexes
M_k	The maximum number of containers of type k
M	A large number

Table 6.2. Data for the **SCPf** Model

$x_{i,t}$	Order release quantity for SKU i in time t
$y_{k,t}$	"Overtime" fraction of resource k in time t
$\delta_{i,t}$	Binary indicator of production of SKU i in time t
$I^{+}_{i,t}$	Inventory of SKU i to carry in time t
$I^{-}_{i,t}$	Quantity of SKU i backordered in time t
$X^{-}_{i,t}$	Negative deviation from the last plan
$X^{+}_{i,t}$	Positive deviation from the last plan
$\theta_{k,m,t}$	Ones in order for containers m of type k in time bucket t

Table 6.3. Variables Set by the **SCPf** Model

6.6 Aggregation and Consolidation

The models described to this point will often be far too large to be solved in a reasonable amount of time. The level of detail must be reduced in order to find a solution. The resolution can then be increased.

Minimize:

$$\sum_{t=1}^{T}\Bigg[\sum_{i=1}^{P}\Big(A(i)I^-_{i,t}+H(i)\left(I^+_{i,t}+LT(i)x_{i,t}\right)$$
$$+C(i)\delta_{i,t}+V(i)x_{i,t}+E(i,t)\left(X^+_{i,t}+X^-_{i,t}\right)\Big)$$
$$+\sum_{k=1}^{K}O(k,t)y_{k,t}+\sum_{k\in\mathcal{K}}\sum_{m=1}^{M_k}B(k,m)\theta_{k,m,t}\Bigg] \qquad \textbf{(SCPf)}$$

subject to:

$$\begin{array}{lll}
I_{i,t}(x,\delta)+\sum_{\tau=1}^{t}D(i,\tau) & \geq 0 & i=1,\ldots,P,\ t=1,\ldots,T\\
\sum_{i=1}^{P}[U(i,k)x_{i,t}+S(i,k)\delta_{i,t}] & \leq 1+y_{k,t} & t=1,\ldots,T,\ k=1,\ldots,K\\
y_{k,t} & \leq F(k,t) & t=1,\ldots,T,\ k=1,\ldots,K\\
\delta_{i,t} & \geq \frac{x_{i,t}}{M} & i=1,\ldots,P,\ t=1,\ldots,T\\
\delta_{i,t} & \in\{0,1\} & i=1,\ldots,P,\ t=1,\ldots,T\\
y_{k,t} & \geq 0 & t=1,\ldots,T,\ k=1,\ldots,K\\
x_{\ell,t} & \geq 0 & \forall\, t,\ell\in\bigcup_{i=1}^{P}\mathcal{L}(i)\\
x_{\ell,t} & = 0 & \forall\, t,\ell\notin\bigcup_{i=1}^{P}\mathcal{L}(i)\\
I^+_{\ell,t} & = 0 & \forall\, t,\ell\in\bigcup_{i=1}^{P}\mathcal{L}(i)\\
I^-_{\ell,t} & = 0 & \forall\, t,\ell\in\bigcup_{i=1}^{P}\mathcal{L}(i)\\
I^+_{\ell,t} & \geq 0 & \forall\, t,\ell\notin\bigcup_{i=1}^{P}\mathcal{L}(i)\\
I^-_{\ell,t} & \geq 0 & \forall\, t,\ell\notin\bigcup_{i=1}^{P}\mathcal{L}(i)\\
I^+_{\ell,t}-I^-_{\ell,t} & = I_{\ell,t}(x,\delta) & \forall\, t,\ell\notin\bigcup_{i=1}^{P}\mathcal{L}(i)\\
I_{i,T}(x,\delta) & \geq I(i,T+1) & i=1,\ldots,P\\
X^+_{i,t}-X^-_{i,t} & = x_{i,t}-X(i,t) & i=1,\ldots,P,\ t=1,\ldots,T\\
X^+_{i,t} & \geq 0 & i=1,\ldots,P,\ t=1,\ldots,T\\
X^-_{i,t} & \geq 0 & i=1,\ldots,P,\ t=1,\ldots,T\\
\sum_{i=1}^{P}U(i,k)x_{i,t} & \leq \sum_{m=1}^{M_k}\theta_{k,m,t} & k\in\mathcal{K},\ t=1,\ldots,T\\
\theta_{k,m,t} & \geq \theta_{k,m+1,t} & \forall\, t,k\in\mathcal{K},m=1,\ldots,M_k-1\\
\theta_{k,m,t} & \in\{0,1\} & \forall\, t,k\in\mathcal{K},m=1,\ldots,M_k
\end{array}$$

Fig. 6.2. SCPf Model

We will work with the concept of *aggregation* and *disaggregation.* Products and resources are grouped together and considered as single entities. Once a plan is found in terms of aggregated entities, a good plan with more detail

$$I_{i,t}(x,\delta) \equiv \sum_{\ell \in \mathcal{L}(i)} \sum_{\tau=1}^{t-LT(\ell)} x_{\ell,\tau} + I(i,0) - \sum_{\tau=1}^{t} D(i,\tau) - \sum_{j=1}^{P} (R(i,j)x_{j,\tau})$$

Fig. 6.3. SCPf Inventory Macro for Master SKUs

$$I_{i,t}(x,\delta) \equiv \sum_{\tau=1}^{t-LT(i)} x_{i,\tau} + I(i,0) - \sum_{\tau=1}^{t} \left(D(i,\tau) + \sum_{j=1}^{P} (R(i,j)x_{j,\tau} + W(i,j)\delta_{j,\tau}) \right)$$

Fig. 6.4. SCPf Inventory Macro for SKUs that are not Master SKUs

can be created that corresponds to it. As with all modeling, decisions about how to group parts and resources is part art and part science.

6.6.1 Consolidating Resources

An important concept for consolidating resources is the notion of a *bottleneck*. The word is so overused that its meaning has become somewhat blurred, but typically serves to signify a limiting factor. Consider an extreme example. Suppose that a supply chain consists entirely of three resources and every SKU is routed through them in order. If for all SKUs, the first and third servers are always much faster and more reliable than the second, then the second server would clearly be the bottleneck. For many planning purposes, there would be no need to include resources one and three in the model.

What effect do non-bottleneck resources have on the model? For non-bottleneck resources k, constraints such as

$$\sum_{i=1}^{P} U(i,k)x_{i,t} \leq 1 + y_{k,t}$$
$$y_{k,t} \geq 0$$
$$y_{k,t} \leq F(k,t)$$

will not effect the solutions obtained. The value of $\sum_{i=1}^{P} U(i,k)x_{i,t}$ will always be less than one. The non-bottleneck resources make it harder to solve because there are variables and constraints associated with each resource. But the y variables are always zero and the constraints are never *binding*.

A constraint is said to bind when the left hand side is equal to the right hand side. One can see that a constraint cannot be effecting the solution unless it binds. Consider simple capacity constraints without overtime:

$$\sum_{i=1}^{P} U(i,k)x_{i,t} \leq 1$$

If $\sum_{i=1}^{P} U(i,k)x_{i,t}$ is strictly less than one for some particular resource, k, for all SKUs i and all times t, then some other constraint(s) and/or the objective function are determining the values of $x_{i,t}$ for all i and t. To put it another way: the solution will be the same with and without the constraints

$$\sum_{i=1}^{P} U(i,k)x_{i,t} \leq 1$$

for this particular k.

We will refer to the difference between the left hand side and the right hand side as the *slack*. We can define a bottleneck as a resource with no slack in some or all of its capacity constraints. A non-bottleneck is a resource with slack in all of its capacity constraints. Resources that cannot conceivably become bottlenecks can safely be removed from the models that we have developed and the solutions will not be effected. However, the time required to find a solution will be reduced. The only trick is to know which resources would have no slack if they were in the model.

One possibility is to use intuition, engineering estimates and historical data to drop non-bottlenecks from the model. Typically, the resources for which capacity constraints are binding (or overtime is needed) do change over time as the demand mix evolves. Obviously, the installation of new capacity or removal of resources can also change the location of the bottleneck. One must update the model to reflect changing conditions.

Another method of looking for capacity slack is to execute the model. This can also be used to supplement estimates and intuition. Of course, if the reason to reduce the number of resources in the model is that the model is too big, this can be problematic. Modern modeling languages provide features that can be helpful for trying the model with some resources included and then again with them excluded. In this way one can test suspicions of capacity limitations without the need to solve a model that is excessively large. More resources can be checked at the same time if the model size can be reduced via other means. The model size can also be reduced by aggregating SKUs which we now consider.

6.6.2 Aggregating Parts

Part aggregation is based on the idea of part *families* that share similar production characteristics. We will use the symbol $\mathcal{F}(\ell)$ to refer to part family number ℓ. The set $\mathcal{F}(\ell)$ contains the list of part indexes that are in the family. We define the production quantity for an entire family, ℓ in period t to be $\hat{x}_{\ell,t}$. We could use any of the models that we have developed so far, but the x variables would be replaced by $\hat{x}$ with appropriate adjustments to the data.

Variables. If the variables x are used for the disaggregated plan, it is easy to see that:

$$\hat{x}_{\ell,t} = \sum_{i \in \mathcal{F}(\ell)} x_{i,t}$$

But normally, we will have solved the aggregated model to get values for $\hat{x}$ variables for the aggregate parts and we will want to estimate the corresponding x values for the individual SKUs. For this purpose we will use the data $Pr(i, \ell)$, which give the fraction (or average fraction, or estimated fraction) of family ℓ production that is SKU i production. We can think of the letters Pr as representing either "proportion" or "probability" depending on the context. Once we have these data, and we have obtained values for $\hat{x}$, then we can compute the corresponding values for x using

$$x_{i,t} = Pr(i, \ell)\hat{x}_{\ell,t}.$$

By using data about past production quantities, we can estimate good values for Pr. We use the symbol $Avg(x_i)$ to refer to the average per period production of SKU i and $Avg(\hat{x}_\ell)$ for the average production quantities for the entire family of which SKU i is a member. A good estimate for Pr is then given by

$$Pr(i, \ell) = \frac{Avg(x_i)}{Avg(\hat{x}_\ell)}$$

if $i \in \mathcal{F}(\ell)$, and zero if $i \notin \mathcal{F}(\ell)$. These values can be adjusted if there are reasons to believe that the mixture within a family will change. If there is no historical production data for an SKU, the values of Pr must be based on an educated guess or marketing estimates.

Data. We can use the values of Pr to switch back and forth between aggregated and disaggregated data and construct the requirements data between families. Consider families ℓ and ℓ'. We can write:

$$\hat{U}(\ell, k) = \sum_{i \in \mathcal{F}(\ell)} Pr(i, \ell)U(i, k)$$

$$\hat{R}(\ell, \ell') = \sum_{i \in \mathcal{F}(\ell)} \sum_{j \in \mathcal{F}(\ell')} Pr(j, \ell')R(i, j)$$

and so forth for the other data values, except changeover data. This is intertwined with issues related to part family formation, so it is discussed in the next subsection.

Forming Part Families. Obviously, the quality of the solution found using aggregate SKUs will depend in large measure on the quality of the families. In an organization with thousands of SKUs assignment of parts to families can be a big undertaking. It is non-trivial even with hundreds of SKUs. Many organizations have grouped their parts into families for purposes other than using optimization models for production planning, but these families may work quite well for our purposes. Industrial and manufacturing engineers often create part families to facilitate process planning and computer aided (re)design of parts.

We can provide some guidance on family creation, if they have not already been formed for other purposes. All other things being equal it is better when all members of a family share a routing (perhaps using consolidated resources), all members should have similar changeover characteristics for bottleneck resources, and it is desirable for all members to have components from as few different families as possible and/or to be components in as few different families as possible. We now discuss these rules in a little bit more detail.

If there are alternative routings, then all members of a family should share a routing because the aggregated capacity check might not be very accurate otherwise. In other words, if some family members can use an alternate routing, but others cannot, then the meaning of a solution is not as clear as it would be if all members can use the same routings. This condition can be tested as follows: If i and j are both members of $\mathcal{F}(\ell)$ and $U(i,k) \neq 0$, but $U(j,k) = 0$, then parts i and j are in the same family, but do not have the same routing.

A similar, but more important situation arises with respect to bottleneck changeover characteristics. Two cases must be distinguished:

1. The changeover that is required between members of the same family is very small compared to the changeover between families. This is the prefered situation. It is reasonable under these conditions to assume that items from the same family will be sequenced one after the other. The changeover time to a member of the family can be used as the value for S. In fact, using aggregated parts will probably produce a more accurate portrayal of the changeover effort than keeping the SKUs separate.
2. The changeover effort within the family is significant. This is trouble. The correct course of action depends on the purpose of the model and the severity of the problem. For rough cut work, one can drop the delta variables and the changeover data and instead just add average changeover times to the capacity utilization data, U. An alternative is to look at historical data and see roughly how much capacity is usually used for this family and use that for S if it is reasonably stable.

A more subtle issue is the desirability of shared components and shared uses. Although the model will still be valid, it is easier to solve if the matrix of R values is mostly zeros, which will almost always be the case before aggregation. If the families are formed completely at random with respect to the bill of materials, then the resulting $\hat{R}$ matrix will be very dense, which is to say that it will not have very many zeros. This will degrade the solution time. Also, it simply makes sense to form families that occupy a somewhat contiguous region of the bill of materials.

Aggregation Example. To illustrate aggregation, consider the example given in Figure 3.3 (page 22) and §4.1. To make the example easier to read, we use the SKU name as the product index instead of a serial number.

Suppose our goal is to reduce the number of variables to speed up computation. One way to do that would be to group the end items together in one family, a second family would be components that are manufactured in-house, and a third would be the purchased components. That results in the following families:

$$\mathcal{F}(1) = \{\text{AJ8172, TR1777}\}$$
$$\mathcal{F}(2) = \{\text{LQ8811, NN1100}\}$$
$$\mathcal{F}(3) = \{\text{RN0098, RN0099, WN7342}\}$$

Suppose that historically AJ8172 has had three times as much demand and production as TR1777 and that this is expected to continue. This prediction would imply that $Pr(\text{AJ8172}, 1) = 0.75$ and $Pr(\text{TR1777}, 1) = 0.25$.

One way to acquire the data for the other families is to base it on these estimates. Consider a demand (or production) of 100 end items. So that means the quantity for the first family is 100, and we can disaggregate as follows: $0.75 \times 100 = 75$ of them will be AJ8172 and $0.25 \times 100 = 25$ will be TR1777. A simple bill of materials explosion results in the following component quantities for the second family:

LQ8811: $2 \times 75 = 150$
NN1100: $1 \times 150 = 150$.

Hence, $Pr(\text{LQ8811}, 2) = 0.5$ and $Pr(\text{NN1100}, 2) = 0.5$ since they are equal parts of the total for the family. For the purchased components, we would have

RN0098: $1 \times 75 + 3 \times 25 = 150$
RN0099: $1 \times 25 = 25$
WN7342: $1 \times 150 = 150$,

which results in a total family demand of 325. This implies a fraction $Pr(\text{RN0098}, 3) = 150/325 = 0.46$, which is the same as $Pr(\text{WN7342}, 3)$. The fraction of the third family devoted to RN0099 is fairly small: $Pr(\text{RN0099}, 3) = 0.08$.

To continue our exercise of the notation, we compute the aggregate, direct requirements for purchased components to create an end item:

$$\hat{R}(3, 1) = \sum_{i \in \mathcal{F}(3)} \sum_{j \in \mathcal{F}(1)} Pr(j, 1) R(i, j),$$

which, after factoring the Pr terms, is

$$\begin{aligned}
&Pr(\text{AJ8172}, 1)\times \\
&(R(\text{RN0098}, \text{AJ8172}) + R(\text{RN0099}, \text{AJ8172}) + R(\text{WN7342}, \text{AJ8172})) \\
&+Pr(\text{TR1777}, 1)\times \\
&(R(\text{RN0098}, \text{TR1777}) + R(\text{RN0099}, \text{TR1777}) + R(\text{WN7342}, \text{TR1777})).
\end{aligned}$$

Remember that $R(i,j)$ refers to the direct requirements for SKU i to make one j, consequently, many of the values are zero:

$$\hat{R}(3,1) = 0.75 \times (1 + 0 + 0) + 0.25 \times (3 + 1 + 0) = 1.75.$$

Similar calculations provide values for $\hat{R}(2,1)$ and $\hat{R}(3,2)$.

In the example given in §4.1 there were two resources: HR-101 and MT-402. For the sake of illustration, suppose that HR-101 is not considered to be a bottleneck and therefore will not be included in the aggregate planning process. That leaves us with only one resource to consider. Since LQ8811, which is in family 2, is the only SKU that uses MT-402, the only non-zero value of $\hat{U}$ will be $U(\hat{2},1) = 0.5 \times \frac{1}{300}$.

Aggregate demand data is often obtained from marketing estimates given in terms of anticipated revenues for a market segment. Segment demand estimates are converted to family demand estimates using simple ratios. For example, suppose that AJ8172 and TR1777 constitute 40% of the sales volume for a particular market segment. If we know that they have an average sales price of \$130 each and that the estimated demand for this segment in period 27 is \$50000, then we can compute $\hat{D}(1,27) = 0.4 \times 50000/130 = 154$.

Values for $\hat{D}$, $\hat{U}$, and $\hat{R}$ can be used to construct an aggregate version of the model **MRPII** as shown in Figure 6.5. The values for $\hat{LT}$ can either be a weighted average, as is done for the other data elements, or they can be the largest value within their family, which is more conservative.

$$\text{minimize: } \sum_{i=1}^{P}\sum_{t=1}^{T}(T-t)\hat{x}_{i,t}$$

subject to:

$$\sum_{\tau=1}^{t-\hat{LT}(i)} \hat{x}_{i,\tau} + \hat{I}(i,0) - \sum_{\tau=1}^{t}\left(\hat{D}(i,\tau) + \sum_{j=1}^{3}\hat{R}(i,j)\hat{x}_{j,\tau}\right) \geq 0 \qquad i = 1,\ldots,3,\ t = 1,\ldots,T$$

$$\sum_{i=1}^{3}\hat{U}(i,1)\hat{x}_{i,t} \leq 1 \qquad t = 1,\ldots,T$$

$$\hat{x}_{i,t} \geq 0 \qquad i = 1,\ldots,3,\ t = 1,\ldots,T$$

Fig. 6.5. Aggregate Version of the **MRPII** Model For the Three Family, One Resource Example

The $\hat{x}$ values from an aggregate **MRPII** model can be disaggregated to give rough estimates of production quantities. Another use of an aggregated data is to implement an aggregate version of **SCPc** and then examine the "overtime" variables y_{kt} for the purpose of anticipating resource needs.

6.6.3 Discussion of Disaggregation

Once we have solved a model based on aggregated parts, the disaggregation relationship

$$x_{i,t} = Pr(i, \ell)\hat{x}_{\ell,t}$$

allows us to estimate production quantities from optimized family production quantities. These can then be used in the same fashion as classic MRP II to check capacity constraints. In other words, the capacity utilization is accumulated at resources along the routing for each SKU. These estimated utilizations can then be compared with available capacity. This can be called a rough cut capacity check. The calculations require no optimization, just the formula $\sum_{i=1}^{P} U(i, k)x_{i,t}$ although it is often preferable to convert the fractional utilizations back into "natural" units for the resources such as tons, square feet, pieces, etc.

As the name implies, for the purpose of rough cut capacity planning, this can be very valuable. Rough cut capacity planning is done in situations where there is some ability to add (or remove) capacity in the near and intermediate term. By using rough cut capacity planning, one can plan ahead. One can make preparations to add capacity where it is likely to be needed based on the estimated utilizations.

For the purpose of production planning in the supply chain rough cut capacity checks are also valuable. The model and the capacity estimates can be refined for those resources whose capacity constraints are likely to effect the solution, namely those resources for which overtime is planned and those for which there is little or no slack in the capacity constraint.

A final potential use of the disaggregated plan is that it can be used as a starting point for the search to the problem of finding a plan based on an aggregation that is not as course. By "not as course" we mean an aggregation that makes use of more families and, therefore, has fewer SKUs assigned to each family. One can proceed through successive reduction in the degree of aggregation, perhaps terminating when each family has only one SKU (which corresponds to the original non-aggregated problem). For each successive reduction in aggregation, the disaggregated solution from the previous problem can be aggregated and used as a starting point for solution.

This idea of systematically increasing the resolution of the planning model can be continued. The solution to the planning problem can be used as a starting point for an aggregated scheduling problem. The solution to such a problem, in turn, can be used for increasingly refined models until a detailed schedule is produced. However, our interest is in planning so we mention this primarily to illustrate the notion of disaggregation and systematically increasing resolution carried to the extreme.

7. Implementation Examples

The models developed in this book can be translated more or less directly into computer languages that have been developed for optimization modeling. We provide implementations of the first three models, **mrp**, **MRPII**, and **SCPc**, using some popular modeling languages: AMPL, GAMS, MPL, OPL, and Mosel. Implementations of additional models and information about additional modeling languages are available on the authors' web site, which is `http://faculty.gsm.ucdavis.edu/~dlw/scm.html`.

This chapter is not intended to be a tutorial on any of the modeling languages. Each of the languages is supported by extensive documentation and training. However, these sample model implementations may prove useful to people interested in using modeling languages for production planning and supply chain management. Used in conjunction with training or documentation, they can provide a jumping off point for creating more sophisticated models in the corresponding languages. Each section in this chapter describes the implementations for a particular language. There is considerable redundancy between the sections because the languages share many features and our goal is to make each description stand alone so that a reader interested in a particular language would not need to refer to the sections describing the others.

The models can also be a useful communication vehicle for experienced optimization modelers working on problems in production planning. They are explained in the main chapters using a writing style that is directed toward production planners. The implementations of the models that are given here can be used to give production planners an introduction to an optimization modeling language.

The implementations are also not intended to serve as a basis for comparing the modeling languages. For one thing, our knowledge and the level of help we received in creating the implementations and the skill level of the people who helped us varied. Furthermore, different concepts are emphasized in each of the implementations. Taken as a group, they provide a nice overview of modeling language capabilities.

All of the modeling languages allow for the use of long, meaningful variable names, which is consistent with accepted good practice among computer programming professionals. In the text we use names such as x and U for variables and data elements so that the formulations can be given in a very compact form, which makes it easier to see the entire model on one page. Some of our implementations retain the use of single character variable names to emphasize the connections with the models in the text. Other implementations make use of long variable and data names.

The languages allow for separation of the model specification from the data and the solver code. For convenience, the data can be in the same file as the model or in a separate file. We have illustrated both forms in our examples. The key thing, though, is that the model need not be changed for new data.

To make the implementations concrete, we provide example data. The production, routing and utilization data come from §3.1 and §4.1. Additional cost data are needed for the **SCPc** model and the more advanced models. We use the following:

Data for the SCPc Examples

1. AJ8172:
   Nominal Production Lead Time: 2 days
   Components: 2 LQ8811, 1 RN0098
   Wasted in Changeover: 10 LQ8811
   External Demand for eight periods: 20, 30, 10, 20, 30, 20, 30, 40

2. LQ8811:
   Nominal Production Lead Time: 3 days
   Components: 1 NN1100, 1 WN7342
   Wasted in Changeover: 10 LQ8811
   No External Demand

3. RN0098:
   Order Lead Time: 4 days
   Components: N/A
   No External Demand

4. NN1100:
   Nominal Production Lead Time: 1 day
   Components: N/A
   No External Demand

5. WN7342:
   Order Lead Time: 12 days
   Components: N/A
   No External Demand

We use the words “Nominal Lead Time” and “Order Lead Time” in this table to refer to the production lead times that will be used for planning as opposed to “actual” lead times, which can vary. We make use of an inventory tolerance constraint as described on page 65 that requires ending inventory to be within a certain range, say up to 25% change ($IT = 0.25$), of the beginning

SKU	$U(i,k)$ AJ8172	LQ8811	NN1100	$S(i,k)$ AJ8172	LQ8811	NN1100
HR-101	1/480	1/960	0.000001	0.2	0.2	0
MT-402		1/300			0.5	

Table 7.1. Resource Oriented Data for the Implementation of the Models

inventory. The example in §4.1 provides capacity utilization data for SKUs AJ8172 and LQ8811, which we use in the implementations described here; see Table 7.1. The utilization of resource HR-101 by SKU NN1100 is taken to be 0.000001 in the example data files. The maximum fraction of resource k that can be added in t (i.e., $F(k,t)$) is set to one half for all pairs (k,t). The marginal costs $O(k,t)$ in dollar per capacity fraction added to resource k is assumed to be equal to one for all pairs (k,t). Finally, Table 7.2 summarizes data for the SKUs involved in the models.

SKU	AJ8172	LQ8811	RN0098	NN1100	WN7342
$LT(i)$	2	3	4	1	12
$I(i,0)$	90	300	100	0	900
$LS(i)$	100	400	100	1	1000
$C(i)$	800	4200	1	1	1
$H(i)$	2	1	0.5	0.1	0.1
$A(i)$	400	100	4	4	4

Table 7.2. SKU Oriented Data for the Implementation of the Models

7.1 AMPL

A computer program is available that implements the modeling language AMPL. The acronym stands for "Algebraic Modeling Language for Mathematical Programming." This language (or, more accurately, its implementation as software) is useful as an interface to various solvers for mathematical programming (e.g., CPLEX or Xpress-MP). The AMPL homepage, `http://www.ampl.com/`, contains additional information and facilities for trying small instances.

The specification of a model is done using a model file that contains an algebraic description of the model in a form very similar to that used in this book. To create a problem instance, a model file is combined with a data file. Data files provide the input data for a particular instance. The following subsections contain model files for the basic models that we have presented along with sample data files.

Before proceeding with the models that we have implemented using the modeling language, a few words of introduction concerning the structure of an AMPL model are in order. AMPL models are given by statements, each terminated by a semicolon. All text after a sharp sign, #, is considered to be a comment and is ignored by the language processor. AMPL models have the following components:

- *Sets* — In addition to the straightforward use of this component to define sets, it is also used in a more subtle way to provide names instead of numbers for indexes. One can create a set of product names and then have summations across all members of the set. The line

  ```
  set PP ordered;          # Set of SKU Numbers
  ```

  declares the existence of a set that will hold the names of the products. It is declared to be an ordered set so that it can be used as an index. The names of the products are read from a data file when the AMPL software runs. The advantage of using a set for indexes is that reports are more meaningful when production is given for an SKU such as LQ8811 rather than product number 2. A similar line

  ```
  set TT ordered;          # Set of Time Buckets
  ```

  is used to declare the set of time buckets.

- *Parameters* — Most of the data elements for the model (those that are not sets) are given as parameters. Parameters can be given values in the model file, but the values are usually read from a separate data file. The line

  ```
  param LT {PP} integer; # Lead Time
  ```

  establishes the parameter `LT` which is the list of lead times. The argument `PP` is declared to be the set of product indexes earlier in the model file.

- *Variables* — This component gives the names of the variables that will be given values by the solver. For example, the statements

  ```
  var d {PP, TT} binary; # Production indicator
  var x {PP, TT} >=0;    # Number of each SKU to order
  ```

  declare all of the variables needed for the model **mrp**. Both `d` and `x` are indexed over the products and the time buckets. The variable named `d` is used for the variable that we have called δ and is declared to be binary because $\delta \in \{0, 1\}$. Simple bounds, such as ≥ 0 can be given when variables are declared rather than in the constraints. This makes the model a little easier to read and perhaps provides some solver efficiency so `x` is declared to be `>=0`.

- *Objective Function* — The objective function indicates the goal, e.g., minimization as well as the function written in terms of the parameters, sets, and variables that have been defined. For example, the statement

  ```
  minimize objective: sum {i in PP, t in TT}
                          (T-ord(t)) * x[i,t];
  ```

  gives the objective function for the model **mrp**. In this expression, the AMPL syntax is illustrated by the replacement of

  $$\sum_{i=1}^{P}\sum_{t=1}^{T}$$

  with `sum {i in PP, t in TT}` and $x_{i,t}$ with `x[i,t]`. Notice that we use `(T - ord(t))` rather than $T - t$. We use `ord(t)` instead of t because in the AMPL implementation we choose to have the times be an ordered set. In this expression, we want the ordinal number of the time ("ord" is short for "ordinal") rather than the name of a corresponding time bucket that is contained in the set of times.

- *Constraints* — Constraints are expressed as functions of the parameters, sets and variables that have been defined. Constraints are constructed using operators such as

  ```
  <, >, <=, >=.
  ```

  For example, the line

  ```
  subject to LotSize {i in PP, t in TT}:
     x[i,t] - d[i,t]*LS[i] >= 0;
  ```

  gives the lotsizing constraint for the **mrp** model. Having more than one constraint may result in all constraints in one block after the `subject to` statement or each constraint given separately after a `subject to` statement.

These components are combined to create a model. This very brief introduction should be enough to understand the models that follow.

7.1.1 mrp Model

An AMPL implementation of **mrp** (see Figure 3.7) is as follows:

```
set PP ordered;                  # Set of SKU Numbers
set TT ordered;                  # Set of Time Buckets

param P integer := card(PP); # Number of SKUs
param T integer := card(TT); # Number of Time Buckets
param M  >= 0;                   # Large Number
param LT {PP} integer;           # Lead Time
param R {PP, PP} integer;        # Number of i to make one j
param D {PP, TT} integer;        # External Demand for an item ...
                                 # ... in a period
param I {PP} integer;            # Beginning Inventory
param LS {PP} integer;           # Lot Size

var d {PP, TT} binary;           # Production indicator
var x {PP, TT} >= 0;             # Number of each SKU to order

# -----------------------------------------------------------------

minimize objective:
   sum {i in PP, t in TT} (T-ord(t)+1) * x[i,t];

# -----------------------------------------------------------------

subject to MaterialRequirement {i in PP, t in TT}:
   (sum {s in TT: ord(s) <= ord(t)-LT[i]} x[i,s]) + I[i]
   - sum {s in TT: ord(s) <= ord(t)}
        (D[i,s] + sum {j in PP} R[i,j] * x[j,s])
   >= 0;

subject to LotSize {i in PP, t in TT}:
   x[i,t] - d[i,t]*LS[i] >= 0;

subject to ProductionIndicator {i in PP, t in TT}:
   d[i,t] - x[i,t]/M >= 0;
```

In the first part the sets, parameters and variables are declared. Comments at the end of the line indicate the function of each letter. The notation `P integer := card(PP)` indicates that P is to be an integer that takes on the value that is the number of elements in the set PP ("card" is an abbreviation for "cardinality").

The objective function is the content of the second segment, the constraints are defined in the last section. The objective function is a little bit different from the function in the abstract model:

```
minimize objective:
   sum {i in PP, t in TT} (T-ord(t)+1) * x[i,t];
```

First, we use `T-ord(t)` instead of $T - t$ because we choose to represent the times as a set of names. This makes the reports easier to read, but means that in order to have the bucket number that corresponds to time t, we need to use `ord(t)`. If we simply let t be an integer, we would not have had to use the `ord` function, but reports would have had numbers for each period instead of meaningful names for the time periods. This is particularly problematic in practice where the reports must be generated each week and two reports where period 1 means two different things can be very confusing. Hence, it is better to use meaningful period names as we have done.

The other difference is due to solver issues. To avoid production quantities in the last time bucket that will not actually be completed before the end of the time horizon covered by the model we insert "+1," which insures that production is always penalized a little bit. Without this penalty, production in period T has no cost in the objective function but as part of the search for an optimal solution, the solver might assign some production to this period. Production ordered for the last period will generally not help or hurt anything in the model because it will not be completed before the end of the horizon. However, spurious production quantities look bad on reports, so this little trick is useful. In this case, we could add any number we wanted and the optimal values for the variables would not be affected (the value of the objective function itself would be different; however, the objective function value for **mrp** has no quantitative economic meaning).

7.1.2 mrp Data

Now the model can be used for our example data. We take the data from the bill of materials, master production schedule and the inventory status files we presented in §3.1. First the list of SKUs, their lead times, and the set of time buckets have to be defined. The beginning inventory and the minimum lot size that depend on the SKU can be also entered as a list of values. The number of SKU i needed to produce one j and the external demand of every SKU at a time period t must be provided. In our case all components indicate no external demand, only the end item AJ8172 has values for every time bucket. We chose 10000 as the value for the "big number," `M`, in our example.

```
param M := 10000;                  # Large number

param: PP :   LT :=                # Items with Lead Times
       AJ8172 2
       LQ8811 3
       RN0098 4
       NN1100 1
       WN7342 12;

set TT := 1jan04                   # Time Buckets
          2jan04
          3jan04
          4jan04
```

```
             5jan04
             6jan04
             7jan04
             8jan04;

param R :                              # Number of i to produce one j
         AJ8172 LQ8811 RN0098 NN1100 WN7342  :=
AJ8172   0      0      0      0      0
LQ8811   2      0      0      0      0
RN0098   1      0      0      0      0
NN1100   0      1      0      0      0
WN7342   0      1      0      0      0;

param D :                              # external demand for i in t
         1jan04 2jan04 3jan04 4jan04 5jan04 6jan04 7jan04 8jan04 :=
AJ8172   20     30     10     20     30     20     30     40
LQ8811   0      0      0      0      0      0      0      0
RN0098   0      0      0      0      0      0      0      0
NN1100   0      0      0      0      0      0      0      0
WN7342   0      0      0      0      0      0      0      0;

param I := AJ8172 90                   # Beginning Inventory of SKU i
           LQ8811 300
           RN0098 100
           NN1100 0
           WN7342 900;

param LS := AJ8172 100                 # Lot Size
            LQ8811 400
            RN0098 100
            NN1100 1
            WN7342 1000;
```

7.1.3 Results of Running mrp

To get the results of our optimization model we can use a batch file. The batch file mrp.run loads the model mrp.mod and the data from mrp.dat, runs the problem, and displays an optimal solution of the variables x and d, and some additional text lines.

The batch file includes AMPL commands to carry out and direct the solution process. The following batch file executes the model using the associated data file:

```
model mrp.mod;
data mrp.dat;
solve;

print "";
print "little mrp with 5 SKUs and 8 time buckets";
display x;
display d;
```

Execution results in an objective function value of 6800 and a material requirements plan for the production quantity at every day which is shown by the variable x[i,t].

```
little mrp with 5 SKUs and 8 time buckets
x [*,*]
:        AJ8172 LQ8811 RN0098 NN1100 WN7342    :=
1jan04      0      0      0      0      0
2jan04      0      0    100    400      0
3jan04    100    400      0      0      0
4jan04      0      0      0      0      0
5jan04      0      0      0      0      0
6jan04    100      0      0      0      0
7jan04      0      0      0      0      0
8jan04      0      0      0      0      0
;

d [*,*]
:        AJ8172 LQ8811 RN0098 NN1100 WN7342    :=
1jan04     0      0      0      0      0
2jan04     0      0      1      1      0
3jan04     1      1      0      0      0
4jan04     0      0      0      0      0
5jan04     0      0      0      0      0
6jan04     1      0      0      0      0
7jan04     0      0      0      0      0
8jan04     0      0      0      0      0
;
```

The values of d are printed here in the interest of completeness. They are redundant in that they simply reflect the non-zero entries in the production orders implied by the x values.

7.1.4 MRPII Model

To implement the model **MRPII** as shown in Figure 4.2 one must add the new parameters, sets, and constraints to the model for **mrp** and drop the lot sizing constraint. A set of resources has to be added and the values for the fraction of each resource needed by one SKU i have to be provided as data. An **MRPII** model looks like the following:

```
set PP ordered;                     # Set of SKU Numbers
set TT ordered;                     # Set of Time Buckets
set KK ordered;                     # Set of Resources

param P integer := card(PP); # Number of SKUs
param T integer := card(TT); # Number of Time Buckets
param K integer := card(KK); # Number of Resources

param I {PP} integer;               # Beginning Inventory
param LT {PP} integer;              # Lead Time
```

```
param R {PP, PP} integer;     # Number of SKUs i to make one SKU j
param D {PP, TT} integer;     # Ext. Demand for an item in a period
param U {PP, KK};             # Frac. of Res. k needed by one SKU i

var x {PP, TT} >=0;           # Number of SKUs to produce

# -----------------------------------------------------------------

minimize objective:
   sum {i in PP, t in TT} (T-ord(t)+1) * x[i,t];

# -----------------------------------------------------------------

subject to MaterialRequirement {i in PP, t in TT}:
   (sum {s in TT: ord(s) <= ord(t)-LT[i]} x[i,s]) + I[i]
   - sum {s in TT: ord(s) <= ord(t)}
         (D[i,s] + sum {j in PP} R[i,j] * x[j,s])
   >= 0;

subject to Capacity {t in TT, k in KK}:
   sum {i in PP} U[i,k] * x[i,t] <= 1;
```

As we did in "little-mrp" we have to insert "+1" into the objective function again to avoid meaningless production quantities in the last time bucket.

7.1.5 Data for MRPII

In addition to the data provided for **mrp**, the parameters $U(i,k)$ and the list of resources, K, must be specified. The minimum lot sizes that we used for **mrp** are not part of this model.

```
param: PP :   LT :=                # Items with Lead Times
       AJ8172 2
       LQ8811 3
       RN0098 4
       NN1100 1
       WN7342 12;


set TT := 1jan04                   # Time Buckets
          2jan04
          3jan04
          4jan04
          5jan04
          6jan04
          7jan04
          8jan04;

set KK := HR-101                   # Resources
          MT-402;
```

```
param R :                              # Number of i to produce one j
        AJ8172 LQ8811 RN0098 NN1100 WN7342  :=
AJ8172  0      0      0      0      0
LQ8811  2      0      0      0      0
RN0098  1      0      0      0      0
NN1100  0      1      0      0      0
WN7342  0      1      0      0      0;

param D :                              # external demand for i in t
        1jan04 2jan04 3jan04 4jan04 5jan04 6jan04 7jan04 8jan04 :=
AJ8172  20     30     10     20     30     20     30     40
LQ8811  0      0      0      0      0      0      0      0
RN0098  0      0      0      0      0      0      0      0
NN1100  0      0      0      0      0      0      0      0
WN7342  0      0      0      0      0      0      0      0;

param I := AJ8172 90                   # Beginning Inventory of SKU i
           LQ8811 300
           RN0098 100
           NN1100 0
           WN7342 900;

param U :                          # fraction of resource k needed by one i
        HR-101      MT-402   :=
AJ8172  0.00208333  0
LQ8811  0.00104166  0.00333333
RN0098  0           0
NN1100  0.000001    0
WN7342  0           0;
```

7.1.6 SCPc Model

To build up the **SCPc** model as shown in Figure 5.1 several things are needed that are similar to the artificial "+1" that was inserted into the objective function for **mrp**. To avoid small production quantities at the last time buckets we add a new "Sunset" constraint. This constraint ensures there will be no production quantities for periods that would result in completion after the time horizon consisting of T time buckets. Such production quantities do not incur any cost penalty in the objective function, so although they also do no good, the solver might set some quantities during the search for an optimal solution. To avoid this nuisance, we add a constraint that would be

$$\sum_{\tau=T-LT(i)+1}^{T} x_{i,\tau} = 0 \qquad i = 1, \ldots, P$$

in the **SCPc** model.

```
subject to Sunset {i in PP}:
   sum {s in TT: ord(s) >= T-LT[i]+1}  x[i,s] = 0;
```

Because the inventory after the last time bucket does not result in any costs we add another constraint to bring the inventory to normal levels at the end of the planning horizon as discussed on page 65. The constraint makes use of the data element IT, which stands for inventory tolerance:

```
subject to InventoryTolerance {i in PP}:
   (1-IT)*I[i] <= iplus[i,last(TT)] <= (1+IT)*I[i];
```

The function call `last(TT)` returns the last index in the ordered set of time buckets.

Due to the nature of the constraints, it is useful to have a non-zero (although perhaps very small) value for all $C(i)$. The reason is that the material requirements constraints and the inventory constraints are "one-sided" in the sense that production of components is required for assembly and production is required in order to create inventory, but there is nothing to prevent spurious non-zero production quantities that are not subsequently in inventory or used as components. This can be repaired by adding constraints, but we think a better thing to do is to continue in the spirit of inserting "+1" into the objective function in **mrp**. We simply make sure that there is a small penalty for all production, so that any production that appears in the plan is there for a reason and is not simply an artifact of the computer's solution process.

The model **SCPc** implemented in AMPL looks like the following:

```
# SCPc.mod
# model SCPc

set PP ordered;                 # Set of SKU Numbers
set TT ordered;                 # Set of Time Buckets
set KK ordered;                 # Set of Resources

param P integer := card(PP);    # Number of SKUs
param T integer := card(TT);    # Number of Time Buckets
param K integer := card(KK);    # Number of Resources
param M >= 0;                   # Large Number
param IT <=1, >=0;              # Inventory Tolerance at the end

param LT {PP} integer;      # Lead Time
param R {PP, PP} integer;   # Number of SKUs i to make one SKU j
param D {PP, TT} integer;   # Ext. Demand for an item in a period
param I {PP} integer;       # Beginning Inventory
param U {PP, KK};           # Fraction of res. k needed by one SKU
param F {KK, TT};           # Max frac. of k that can be added in t
param S {PP, KK};           # Frac. of k used to changeover to SKU i
param W {PP, PP};           # Waste of i to setup changeover to j
param H {PP};               # Per period holding cost for SKU i
param C {PP};               # Total changeover cost for SKU i
param O {KK, TT};           # Marginal cost per fraction added
param A {PP};               # Per Period tardiness cost for ext. d.

var d {PP, TT} binary;      # Production indicator
```

```
var x {PP, TT} >=0;          # Number of SKUs to produce
var y {KK, TT}  >=0;         # "Overtime" fraction of res. k in t
var iplus {PP, TT} >=0;      # Inventory of SKU i to carry in t
var iminus {PP, TT} >=0;     # Quantity of SKU i backordered in t

# ---------------------------------------------------------------

minimize objective:
 sum {t in TT}
     (sum {i in PP}
         (A[i]*iminus[i,t] + H[i]*iplus[i,t]
         + C[i]*d[i,t])
     + sum {k in KK} O[k,t]*y[k,t]);

# ---------------------------------------------------------------

subject to mrp {i in PP, t in TT}:
    sum {s in TT: ord(s) <= ord(t)-LT[i]} x[i,s] + I[i]
    - sum {s in TT: ord(s)<=ord(t)}
          (D[i,s] + sum{j in PP} (R[i,j]*x[j,s] + W[i,j]*d[j,s]))
    + sum {s in TT: ord(s)<=ord(t)} D[i,s]
   >= 0;

subject to Capacity {t in TT, k in KK}:
   sum {i in PP} (U[i,k]*x[i,t] + S[i,k]*d[i,t]) - y[k,t] <= 1;

subject to maxFraction {k in KK, t in TT}:
   y[k,t] - F[k,t] <= 0;

subject to ProductionIndicator {i in PP, t in TT}:
   d[i,t] - x[i,t]/M >=0;

subject to Inventory {i in PP, t in TT}:
   iplus[i,t] - iminus[i,t]
   - (sum {s in TT: ord(s) <= ord(t)-LT[i]} x[i,s] + I[i]
     - sum {s in TT: ord(s)<=ord(t)}
           (D[i,s] + sum{j in PP} (R[i,j]*x[j,s] + W[i,j]*d[j,s]))
     )
   = 0;

subject to Sunset {i in PP}:
   sum {s in TT: ord(s) >= T-LT[i]+1}  x[i,s] = 0;

subject to InventoryTolerance {i in PP}:
   (1-IT)*I[i] <= iplus[i,last(TT)] <= (1+IT)*I[i];
```

Notice that rather than using the inventory macro, we have expanded it. For example, the mrp constraint uses the expression for the macro as given on page 48.

7.1.7 Data for SCPc

Beyond the data provided for **MRPII**, a lot of new data must be added to create the data file for **SCPc**:

```
# data SCPc.dat

param: PP :   LT :=             # Items with Lead Times
       AJ8172 2
       LQ8811 3
       RN0098 4
       NN1100 1
       WN7342 12;

set TT := 1jan04                # Time Buckets
          2jan04
          3jan04
          4jan04
          5jan04
          6jan04
          7jan04
          8jan04;

set KK := HR-101                # Resources
          MT-402;


param M := 10000;               # Large number

param IT := 0.25;               # Inventory Tolerance at the end

param R :
        AJ8172 LQ8811 RN0098 NN1100 WN7342 :=
AJ8172  0      0      0      0      0
LQ8811  2      0      0      0      0
RN0098  1      0      0      0      0
NN1100  0      1      0      0      0
WN7342  0      1      0      0      0;

param D :                      # external demand for i in t
        1jan04 2jan04 3jan04 4jan04 5jan04 6jan04 7jan04 8jan04 :=
AJ8172  20     30     10     20     30     20     30     40
LQ8811  0      0      0      0      0      0      0      0
RN0098  0      0      0      0      0      0      0      0
NN1100  0      0      0      0      0      0      0      0
WN7342  0      0      0      0      0      0      0      0;

param I := AJ8172 90           # Beginning Inventory of SKU i
           LQ8811 300
           RN0098 100
           NN1100 0
           WN7342 900;
```

```
param U :                    # fraction of resource k needed by one i
        HR-101       MT-402 :=
AJ8172  0.00208333   0
LQ8811  0.00104166   0.00333333
RN0098  0            0
NN1100  0.000001     0
WN7342  0            0;

param F :                    # max fraction of k that can be added i
        1jan04 2jan04 3jan04 4jan04 5jan04 6jan04 7jan04 8jan04 :=
MT-402  0.5    0.5    0.5    0.5    0.5    0.5    0.5    0.5
HR-101  0.5    0.5    0.5    0.5    0.5    0.5    0.5    0.5;

param S :                           # Frac. of k used to changeover to i
        HR-101      MT-402 :=
AJ8172  0.2          0
LQ8811  0.2          0.5
RN0098  0            0
NN1100  0            0
WN7342  0            0;

param W :                           # Waste of i to changeover to j
        AJ8172      LQ8811  RN0098  NN1100  WN7342 :=
AJ8172  0           0       0       0       0
LQ8811  10          10      0       0       0
RN0098  0           0       0       0       0
NN1100  0           0       0       0       0
WN7342  0           0       0       0       0;

param H := AJ8172 2                 # per period holding cost for SKU i
           LQ8811 1
           RN0098 0.5
           NN1100 0.1
           WN7342 0.1;


param C := AJ8172 800               # total changeover cost for SKU i
           LQ8811 4200
           RN0098 1
           NN1100 1
           WN7342 1;


param O :                           # marginal cost per fraction added
        1jan04 2jan04 3jan04 4jan04 5jan04 6jan04 7jan04 8jan04 :=
MT-402  1      1      1      1      1      1      1      1
HR-101  1      1      1      1      1      1      1      1;


param A := AJ8172 400               # per period tardiness cost for e.d.
           LQ8811 100
           RN0098 4
           NN1100 4
           WN7342 4;
```

7.2 GAMS

The GAMS modeling language is widely recognized as pioneering the development of software for translation of algebraic optimization models into a form that is accessible to solver software. More information about GAMS can be obtained at `http://www.gams.com` along with pointers to more examples.

Within the GAMS language, components of the model are defined and then combined to make a model. Generally, GAMS statements (which may have multiple parts) end with a semicolon and text placed between syntactic entities is considered to be explanatory text. Lines that begin with an asterisk are comments. The required components are:

- *Sets* — In addition to the straightforward use of this component to define sets, it is also used in a more subtle way to provide names instead of numbers for indexes. One can create a set of product names and then have summations across all members of the set. The lines

```
set PP SKU Numbers  / AJ8172, LQ8811, RN0098, NN1100, WN7342 /
    TT Time Buckets / 1jan04*8jan04 /
    KK Resources / HR-101, MT-402 /;
```

  declare and assign values to all of the sets needed to name the indexes for **mrp** and **MRPII**. In the case of data assignments, an asterisk means "and all values through," provided that the difference in the left and right set element consists of numbers and the left number is smaller than the right. Hence, `1jan04*8jan04` means `1jan04, 2jan04, ..., 7jan04, 8jan04`.

- *Parameters, Tables, Scalars* — Most of the data elements for the model (those that are not sets) are given as parameters, tables, or scalars. They are declared and given their values in the same statement, such as the following:

```
table U(PP,KK) fraction of resource k needed by one i
        HR-101      MT-402
AJ8172  0.00208333  0
LQ8811  0.00104166  0.00333333
RN0098  0           0
NN1100  0.000001    0
WN7342  0           0
;
```

  that declares the utilization and assigns it values. The text "fraction of resource k needed by one i" is treated as explanatory text. The index values for `PP` and `KK` such as AJ8172 and HR-101 must be previously assigned to the sets `PP` and `KK`.

- *Variables* — This component gives the names of the variables that will be given values by the solver. For example, the statements

```
binary variable d(PP,TT)   production indicator
positive variable x(PP,TT) number of SKUs to produce;
```

declare all of the variables needed for the model **mrp**. Both `d` and `x` are indexed over the products and the time buckets. The variable name `d` is used for the variable that we have called δ and is declared to be binary because $\delta \in \{0, 1\}$. Simple bounds, such as ≥ 0 can be given when variables are declared rather than in the constraints for better clarity and perhaps some solver efficiency so `x` is declared to be `positive`, which means $x \geq 0$. Extra text, such as "number of SKUs to produce" is not semantic with respect to the model definition, but can be used in reports as explanatory text. A statement such as

```
variable obj;
```

must be included to declare an unrestricted variable to accept the objective function value.

- *Equations* — Equations (which includes inequalities) are used for the objective function and the constraints. Equations must first be declared and then defined. For example, the statement

```
defcap(TT,KK)        capacity;
```

declares the existence of an equation named `defcap` and indicates that there will be a defcap equation for all members of the sets TT and KK, which is the GAMS implementation of $t = 1, \ldots T, \quad k = 1, \ldots, K$. The word "capacity" is treated as explanatory text, which is more than a comment. Inside the GAMS program, the user has access to these strings. This comes in handy for writing reports, etc. The statement

```
defcap(TT,KK).. sum(PP, U(PP,KK)*x(PP,TT)) =L= 1;
```

defines the constraint

$$\sum_{i=1}^{P} U(i,k)x_{i,t} \leq 1 \qquad t = 1, \ldots, T, \quad k = 1, \ldots, K$$

The name "defcap" can then be used in model definitions. Any model with "defcap" will include the capacity constraint. An equation declaration is indicated by an equation name followed by two periods. The sense of the equation or inequality is given by a letter between equal signs. So =L= means $\leq$, while =E= means $=$. Notice that the syntax for summation is the word `sum` followed by the summation indexes and the summands in parenthesis. An objective function is defined by equating an unrestricted variable with a function of the data and variables, such as

```
defobj.. obj =E= sum((PP,TT), (card(TT)-ord(TT))*x(PP,TT));
```

which implements the **mrp** objective function.

- *Model* — The model statement is the word "model" followed by a list of equation names delimited by slashes. For example, the statement

```
model mrp  /defobj, defreq, deflot, defprod/;
```

defines the model for **mrp** (assuming that the equations listed have been defined).

- *Solve* — The solve command instructs the solver to use a model as a basis for minimizing or maximizing a variable. For example,

  ```
  solve mrp  minimizing obj using mip;
  ```

  instructs the solver to minimize the value of the variable named `obj` using the model named `mrp`. The text "using mip" warns the solver software that some variables are binary (see §8.1).

7.2.1 mrp and MRPII Models

Both the **mrp** and **MRPII** models are in one file because the two models share input data and equations. This also shows the power of modeling languages where one can quickly build various instances of the same base model. Having defined the constraints `defreq`, `deflot`, `defprod`, `defred`, and `defcap` as well as the two objective functions, specification of the two models that use various combinations of these building blocks is simply:

```
model mrp  /defobj, defreq, deflot, defprod/;
model mrp2 /defobj, defreq,                defcap/;
```

Here are a few comments about the implementation:

1. In this implementation, the "0" has been removed from the tables. Sometimes this is referred to as a *sparse* representation.
2. The word "alias" is used as a way to get additional indexes over the same set. For example, `alias (TT,TTp)` established TTp to be used in expressions where the model dictates that the constraint is to be for every time t and within the constraint formula there is a summation over time using the index τ.
3. In GAMS, one uses the dollar sign ($) to mean "such that." For example

   ```
   sum(TTp$(ord(TTp)<=ord(TT)-LT(PP))
   ```

   implements

   $$\sum_{\tau=1}^{t-LT(i)}$$

   with `ord(TTp)` giving the value of τ and `ord(TT)` giving the value of t. When a dollar sign appears at the beginning of a line it introduces a directive that is not really part of the model, such as a title to be used in reports.
4. The statement "`option optcr=0.0;`" instructs the solver to seek an optimal, rather than near optimal, solution.

5. Generally, modeling languages are programmable. For example, calculations based on the data can be performed. In this implementation, the total demand from external and implicit demand is derived by going through the BOM treating it as a tree. The code starts with the root node of the tree (level 0) and goes through the nodes in a breadth-first manner, which accumulates a bound on the total demand for subproducts using the following code:

```
parameter lev(PP) level in the BOM
          TD(PP)  bound on total demand extern plus implicit;
scalar runlev level iteration / 0 /;

* Root node gets level 0, all others get -1
lev(PP)$(sum(PPp,R(PP,PPp))) = -1;
TD(PP)$(lev(PP) = 0) = sum(TT,demand(PP,TT)) + LS(PP);
loop(PP$(lev(PP) = runlev),
  runlev = runlev + 1;
  lev(PPp)$R(PPp,PP) = runlev;
  TD(PPp)$R(PPp,PP) = sum(TT,demand(PPp,TT))
                      + R(PPp,PP)*TD(PP) + LS(PP);
);
```

Having an upper bound on the total demand allows calculation of big M for defprod, which is much better than having just “a big number:”

```
parameter M(PP) big M for equation defprod;
M(PP) = TD(PP);
```

6. GAMS is case insensitive, so the variable “D” was changed to “demand.”
7. To avoid small production quantities in the last time buckets we insert “+1” into the objective function, which insures that production is always penalized a little bit. For additional discussion, see page 87.

The GAMS implementation of **mrp** and **MRPII** is as follows:

```
$Title Materials Requirements Planning (mrp/MRPII) Formulations

set PP SKU Numbers  / AJ8172, LQ8811, RN0098, NN1100, WN7342 /
    TT Time Buckets / 1jan04*8jan04 /
    KK Resources / HR-101, MT-402 /;

alias (TT,TTp)
      (PP,PPp);

table      R(PP,PP) number of i to make one j
         AJ8172 LQ8811 RN0098 NN1100 WN7342

AJ8172
LQ8811  2
RN0098  1
NN1100         1
WN7342         1
;
```

```
table     demand(PP,TT) External Demand for an item in a period
        1jan04 2jan04 3jan04 4jan04 5jan04 6jan04 7jan04 8jan04
AJ8172      20     30     10     20     30     20     30     40
LQ8811
RN0098
NN1100
WN7342
;

parameter lev(PP) Level in the production tree
          TD(PP)  Bound on total demand extern plus implicit;
scalar runlev level iteration / 0 /;

* Root node get level 0, all other get -1
lev(PP)$(sum(PPp,R(PP,PPp))) = -1;
TD(PP)$(lev(PP) = 0) = sum(TT,demand(PP,TT)) + LS(PP);
loop(PP$(lev(PP) = runlev),
  runlev = runlev + 1;
  lev(PPp)$R(PPp,PP) = runlev;
  TD(PPp)$R(PPp,PP) = sum(TT,demand(PPp,TT))
                      + R(PPp,PP)*TD(PP) + LS(PP);
);

parameter LT(PP) Lead Time
          I(PP)  Beginning Inventory
          LS(PP) Lot Size;

table SKUdata
          LT     LS     I
AJ8172     2    100    90
LQ8811     3    400   300
RN0098     4    100   100
NN1100     1      1     0
WN7342    12   1000   900
;

LT(PP) = SKUdata(PP,'LT');
LS(PP) = SKUdata(PP,'LS');
I(PP) = SKUdata(PP,'I');


table U(PP,KK) fraction of resource k needed by one i
        HR-101      MT-402
AJ8172  0.00208333  0
LQ8811  0.00104166  0.00333333
RN0098  0           0
NN1100  0.000001    0
WN7342  0           0
;

parameter M(PP) big M for equation defprod;
M(PP) = TD(PP);
```

```
binary variable d(PP,TT)    production indicator
positive variable x(PP,TT) number of SKUs to produce
variable obj;

equation defobj                objective function
         defreq(PP,TT)         material requirement
         deflot(PP,TT)         lot size
         defprod(PP,TT)        production indicator
         defcap(TT,KK)         capacity;

defobj.. obj =E= sum((PP,TT), (card(TT)-ord(TT)+1)*x(PP,TT));

defreq(PP,TT).. sum(TTp$(ord(TTp)<=ord(TT)-LT(PP)), x(PP,TTp))
                 + I(PP)
                =G= sum(TTp$(ord(TTp)<=ord(TT)), demand(PP,TTp)
                     + sum(PPp, R(PP,PPp)*x(PPp,TTp)));

deflot(PP,TT).. x(PP,TT) =G= d(PP,TT)*LS(PP);

defprod(PP,TT).. x(PP,TT) =L= d(PP,TT)*M(PP);

defcap(TT,KK).. sum(PP, U(PP,KK)*x(PP,TT)) =L= 1;

model mrp  /defobj, defreq, deflot, defprod/;
model mrp2 /defobj, defreq,                 defcap/;

option optcr=0.0;
solve mrp  minimizing obj using mip;
solve mrp2 minimizing obj using lp;
```

7.2.2 SCPc Model

To build up the **SCPc** model as shown in Figure 5.1 several things are needed that are similar to the artificial "+1" that was inserted into the objective function for **mrp**. In order to explain the syntax and annotate the model, we discuss a few of them.

To avoid small production quantities at the last time buckets we add a new "Sunset" constraint. This constraint ensures there will be no production quantities for periods that would result in completion after the time horizon consisting of T time buckets. Such production quantities do not incur any cost penalty in the objective function, so although they also do no good, the solver might set some quantities during the search for an optimal solution. To avoid this nuisance, we add a constraint that would be

$$\sum_{\tau=T-LT(i)+1}^{T} x_{i,\tau} = 0 \qquad i = 1, \ldots, P$$

in the **SCPc** model.

```
x.fx(PP,TTp)$(ord(TTp) > card(TT)-LT(PP)) = 0;
d.fx(PP,TTp)$(ord(TTp) > card(TT)-LT(PP)) = 0;
```

The `fx` suffix indicates that the variable should be fixed and the dollar sign means "such that," so the first constraint translates to "`x(PP,TTp)` should be fixed at 0 for all `PP` and `TTp` such that production that starts in period TTp would result in completion after the time horizon (which is given by `card(TT)`)."

Because the inventory after the last time bucket does not result in any costs we add another constraint to bring the inventory to normal levels at the end of the planning horizon as discussed on page 65. The constraint makes use of the data element `IT`, which stands for inventory tolerance:

```
* inventory tolerance
iplus.up(PP,TTp)$(ord(TTp) = card(TT)) = (1+IT)*I(PP);
iplus.lo(PP,TTp)$(ord(TTp) = card(TT)) = (1-IT)*I(PP);
```

The suffix `up` indicates an upper bound while `lo` refers to a lower bound. The data element `IT` is declared and given a value by the statement

```
scalar IT inventory tolerance at the end / 0.25 /;
```

Due to the nature of the constraints, it is useful to have a non-zero (although perhaps very small) value for all $C(i)$. The reason is that the material requirements constraints and the inventory constraints are "one-sided" in the sense that production of components is required for assembly and production is required in order to create inventory, but there is nothing to prevent spurious non-zero production quantities that are not subsequently in inventory or used as components. This can be repaired by adding constraints, but we think a better thing to do is to continue in the spirit of inserting "+1" into the objective function in **mrp**. We simply make sure that there is a small penalty for all production, so that any production that appears in the plan is there for a reason and is not simply an artifact of the computer's solution process.

The GAMS implementation of **SCPc** is as follows; notice that instead of the inventory macro, an additional variable, `inv` is used in this implementation:

```
$Title (SCPc) Formulation

set PP SKU Numbers  / AJ8172, LQ8811, RN0098, NN1100, WN7342 /
    TT Time Buckets / 1jan04*8jan04 /
    KK Resources / HR-101, MT-402 /;

alias (TT,TTp)
      (PP,PPp);
```

```
table     R(PP,PP) number of i to make one j
       AJ8172 LQ8811 RN0098 NN1100 WN7342
AJ8172
LQ8811  2
RN0098  1
NN1100        1
WN7342        1
;

table     demand(PP,TT) External Demand for an item in a period
       1jan04 2jan04 3jan04 4jan04 5jan04 6jan04 7jan04 8jan04
AJ8172     20     30     10     20     30     20     30     40
LQ8811
RN0098
NN1100
WN7342
;

parameter lev(PP) Level in the production tree
          TD(PP)  Bound on total demand extern plus implicit;
scalar runlev level iteration / 0 /;

* Root node get level 0, all other get -1
lev(PP)$(sum(PPp,R(PP,PPp))) = -1;
TD(PP)$(lev(PP) = 0) = sum(TT,demand(PP,TT)) + LS(PP);
loop(PP$(lev(PP) = runlev),
  runlev = runlev + 1;
  lev(PPp)$R(PPp,PP) = runlev;
  TD(PPp)$R(PPp,PP) = sum(TT,demand(PPp,TT))
                      + R(PPp,PP)*TD(PP) + LS(PP);
);

parameter LT(PP) Lead Time
          I(PP)  Beginning Inventory
          H(PP)  per period holding cost
          C(PP)  total changeover cost
          A(PP)  per period tardiness cost;

table SKUdata
           LT     I     H      C     A
AJ8172      2    90   2.0    800   400
LQ8811      3   300   1.0   4200   100
RN0098      4   100   0.5      1     4
NN1100      1         0.1      1     4
WN7342     12   900   0.1      1     4
;

LT(PP) = SKUdata(PP,'LT');
I(PP)  = SKUdata(PP,'I');
H(PP)  = SKUdata(PP,'H');
C(PP)  = SKUdata(PP,'C');
A(PP)  = SKUdata(PP,'A');

scalar IT inventory tolerance at the end / 0.25 /;
```

```
table U(PP,KK) fraction of resource k needed by one i
        HR-101      MT-402
AJ8172  0.00208333  0
LQ8811  0.00104166  0.00333333
RN0098  0           0
NN1100  0.000001    0
WN7342  0           0
;

table F(KK,TT) max fraction of k that can be added in t
        1jan04 2jan04 3jan04 4jan04 5jan04 6jan04 7jan04 8jan04
MT-402     0.5    0.5    0.5    0.5    0.5    0.5    0.5    0.5
HR-101     0.5    0.5    0.5    0.5    0.5    0.5    0.5    0.5
;

table S(PP,KK) fraction of k used to changeover to i
        HR-101    MT-402
AJ8172  0.2         0
LQ8811  0.2         0.5
RN0098  0           0
NN1100  0           0
WN7342  0           0
;


table W(PP,PP) waste of i to changeover to j
        AJ8172     LQ8811  RN0098  NN1100  WN7342
AJ8172
LQ8811      10          10
RN0098
NN1100
WN7342
;

table O(KK,TT) marginal cost per fraction
        1jan04 2jan04 3jan04 4jan04 5jan04 6jan04 7jan04 8jan04
MT-402       1      1      1      1      1      1      1      1
HR-101       1      1      1      1      1      1      1      1
;

parameter M(PP) big M for equation defprod;
M(PP) = TD(PP);

binary variable   d(PP,TT)       production indicator
positive variable x(PP,TT)       number of SKUs to produce
                  y(KK,TT)       overtime frac. of resource k in t
                  iplus(PP,TT)   inventory of SKU i to carry in t
                  iminus(PP,TT)  quantity of SKU i backordered in t
variable          inv(PP,TT)     intermediate inventory variable
                  obj            objective variable;

equation defobj                objective function
```

```
         defreq(PP,TT)       material requirement
         defprod(PP,TT)      production indicator
         defcap(TT,KK)       capacity
         definv(PP,TT)       inventory macro
         definvsplit(PP,TT) inventory split in carry and backorder;

defobj.. obj =E= sum((PP,TT), A(PP)*iminus(PP,TT)
                  + H(PP)*iplus(PP,TT) + C(PP)*d(PP,TT))
                 + sum((KK,TT), O(KK,TT)*y(KK,TT));

defreq(PP,TT).. inv(PP,TT) + sum(TTp$(ord(TTp)<=ord(TT)),
                             demand(PP,TTp)) =G= 0;

defcap(TT,KK).. sum(PP, U(PP,KK)*x(PP,TT) + S(PP,KK)*d(PP,TT)) =L=
                 1 + y(KK,TT);

defprod(PP,TT).. x(PP,TT) =L= d(PP,TT)*M(PP);

definv(PP,TT).. inv(PP,TT) =E= sum(TTp$(ord(TTp)<=ord(TT)-LT(PP)),
                 x(PP,TTp)) + I(PP) - sum(TTp$(ord(TTp)<=ord(TT)),
                 demand(PP,TTp) + sum(PPp, R(PP,PPp)*x(PPp,TTp)
                 + W(PP,PPp)*d(PPp,TTp)));

definvsplit(PP,TT).. iplus(PP,TT) - iminus(PP,TT) =E= inv(PP,TT);

model SCPc  /defobj, definv, defreq, defcap, defprod, definvsplit/;

* Max fraction
y.up(KK,TT) = F(KK,TT);

* Sunset
x.fx(PP,TTp)$(ord(TTp) > card(TT)-LT(PP)) = 0;
d.fx(PP,TTp)$(ord(TTp) > card(TT)-LT(PP)) = 0;

* inventory tolerance
iplus.up(PP,TTp)$(ord(TTp) = card(TT)) = (1+IT)*I(PP);
iplus.lo(PP,TTp)$(ord(TTp) = card(TT)) = (1-IT)*I(PP);

option optcr=0.0;
solve SCPc  minimizing obj using mip;
```

7.3 Maximal MPL

The MPL modeling language is part of the modeling environment supported by Maximal software. For more information about Maximal and MPL, see `http://www.maximal-usa.com`.

MPL statements end with a semicolon. Any text after an explanation point is treated as a comment. Comments may appear anywhere in the model file. The language makes use of section titles, which generally introduce a group of related statements. The definitions used by our models are grouped using the following sections:

- `Title` — This very short section contains just one statement that gives a title to the model. The title is used to refer to the model in displays such as solution listings. For example, the line

  ```
  TITLE
     mrp;
  ```

  establishes that `mrp` should be used on reports and other references to the model.
- `Index` — This section establishes names for the list indexes. For example,

  ```
  sku := (AJ8172, LQ8811, RN0098, NN1100, WN7342);
  ```

  establishes the list of part indexes. Although we use $i = 1, \ldots, P$ in the text, it makes more sense to use a list of actual part names in an implementation. In practice, such names would come from a database; modeling languages support extraction of such names from external sources. Our examples are small, so in the interest of simplicity we include them directly in the model file. As an aside, we note that the index section can also establish list indexes that are strictly numeric rather than a list of names.
- `Data` — The section gives commands related to obtaining the data needed for the problem. As with the indexes, one usually loads data from external files in large applications. For our small examples, the data are provided in the model declaration file itself. For example,

  ```
  InitialInventory[sku] := (90, 300, 100, 0, 900);
  ```

  declares a data vector named `InitialInventory` that is indexed by `sku`. A vector of values is also assigned by the same statement.
- `Decision Variables` — This section declares the names and indexes for the variables whose values are to be set to optimal values by the solver. For example,

  ```
  DECISION VARIABLES
    PrdIndicator[sku,time];
    Prod[sku,time];
  ```

  declares `PrdIndicator`, which plays the role of δ, and `Prod`, which plays the role of x.

The model part of the file begins with the keyword MODEL and is terminated by the keyword END. Our examples make use of the following model sections:

- `MIN` — This keyword flags an objective function to be minimized (as one would expect, `MAX` indicates a maximization objective). For example,

  ```
  MIN Obj = SUM(sku,time: TimeReverse * Prod);
  ```

  implements an objective function similar to the **mrp** objective function The function `TimeReverse` implements $T-t+1$ as shown below. The text `SUM(sku,time:` establishes that summation will be over the indexes `sku` and `time` so those indexes need not be indicated again for `TimeReverse` and `Prod`.
- `Subject To` — This section makes use of the variables and data vectors to establish the constraints. For example,

  ```
  LotSize[sku,time]:
       Prod >= PrdIndicator * Lotsize;
  ```

  implements

  $$x_{i,t} - \delta_{i,t} LS(i) \geq 0 \quad i = 1, \ldots, P, \quad t = 1, \ldots, T$$

  and establishes `LotSize` as the name of the constraint to be used in reports. The text `LotSize[sku,time]` declares that constraint will be repeated over all values of the index sets `sku` and `time` so those indexes need not be indicated again for `Prod`, `PrdIndicator`, or `Lotsize`.
- `Binary` — This section declares that some variables must be given a value of zero or one.

7.3.1 mrp Model

A few comments about the implementation of **mrp** may help the reader to understand the implementation:

1. The index set `sku2` is a copy of `sku` and is declared to allow for SKU by SKU tables such as the bill of materials, which is called `ProdRequire` in this implementation.
2. The data element `Numbertime` plays the role of T and is computed as the number of elements in the time index set using the MPL function `count`.
3. To avoid small production quantities in the last time buckets we insert "+1" into the `TimeReverse` macro, which insures that production is always penalized a little bit. For additional discussion, see page 87.
4. In this implementation the value of M is coded at 10000 using the data element `LM`.

```
TITLE
  mrp;

INDEX
  sku := (AJ8172, LQ8811, RN0098, NN1100, WN7342);
  sku2 := sku;
  time := (jan1,jan2,jan3,jan4,jan5,jan6,jan7,jan8);

DATA
  Numbertime := count(time);
  InitialInventory[sku] := (90, 300, 100, 0, 900);
  Leadtime[sku]         := (2, 3, 4, 1, 12);
  ProdRequire[sku, sku2] := (0, 0, 0, 0, 0,  ! R(sku, sku2)
                             2, 0, 0, 0, 0,
                             1, 0, 0, 0, 0,
                             0, 1, 0, 0, 0,
                             0, 1, 0, 0, 0);

  ExtDemand[sku,time] := (20, 30, 10, 20, 30, 20, 30, 40,
                           0,  0,  0,  0,  0,  0,  0,  0,
                           0,  0,  0,  0,  0,  0,  0,  0,
                           0,  0,  0,  0,  0,  0,  0,  0,
                           0,  0,  0,  0,  0,  0,  0,  0);

  LM := 10000;
  Lotsize[sku] := (100, 400, 100, 1, 1000);
  TimeReverse[time] := FORMULA(Numbertime - time + 1);

DECISION VARIABLES
  PrdIndicator[sku,time];
  Prod[sku,time];

MODEL
  MIN Obj = SUM(sku,time: TimeReverse * Prod);

SUBJECT TO
  MaterialRequirement[sku,time]:
       SUM(time=1..time-Leadtime: Prod)
     + InitialInventory
    >=
       SUM(time=1..time: ExtDemand
         + SUM(sku2: ProdRequire * Prod[sku:=sku2]));

  LotSize[sku,time]:
       Prod >= PrdIndicator * Lotsize;

  ProductionIndicator[sku,time]:
       PrdIndicator >= Prod / LM;

BINARY
  PrdIndicator;

END
```

The `FORMULA` function in MPL is essentially a macro facility, so

```
TimeReverse[time] := FORMULA(Numbertime - time + 1);
```

would have been written in our notation as $TR_t \equiv (T - t + 1)$, assuming we wanted to use TR as the macro name. It implies that `Numbertime - time + 1` should be substituted for `TimeReverse[time]` in the implementation of **mrp**.

7.3.2 MRPII

To implement the model **MRPII** as shown in Figure 4.2 one must add the new parameters, sets, and constraints to the model for **mrp** and drop the lot sizing constraint. A set of resources has to be added and the values for the fraction of each resource needed by one SKU i have to be provided as data. An MPL implementation of the **MRPII** model is:

```
TITLE
  MRP2;

INDEX
  sku := (AJ8172, LQ8811, RN0098, NN1100, WN7342);
  sku2 := sku;
  time := (jan1,jan2,jan3,jan4,jan5,jan6,jan7,jan8);
  resource := (HR_101,MT_402);

DATA
  Numbertime := count(time);

  InitialInventory[sku] := (90, 300, 100, 0, 900);

  Leadtime[sku]         := (2, 3, 4, 1, 12);

  ProdRequire[sku, sku2] := (0, 0, 0, 0, 0, ! R(sku2, sku)
                             2, 0, 0, 0, 0,
                             1, 0, 0, 0, 0,
                             0, 1, 0, 0, 0,
                             0, 1, 0, 0, 0);

  ExtDemand[sku,time] := (20, 30, 10, 20, 30, 20, 30, 40,
                          0,   0,  0,  0,  0,  0,  0,  0,
                          0,   0,  0,  0,  0,  0,  0,  0,
                          0,   0,  0,  0,  0,  0,  0,  0,
                          0,   0,  0,  0,  0,  0,  0,  0);

  ResourceReq[sku,resource] := (0.00208333, 0,
                                0.00104166, 0.00333333,
                                0,          0,
                                0.000001,   0,
                                0,          0);

  TimeReverse[time] := FORMULA(Numbertime - time + 1);
```

```
DECISION VARIABLES
  Prod[sku,time];

MODEL

  MIN Obj = SUM(sku,time: TimeReverse * Prod);

SUBJECT TO
  MaterialRequirement[sku,time]:
       SUM(time=1..time-Leadtime: Prod)
    + InitialInventory
   >=
       SUM(time=1..time: ExtDemand
         + SUM(sku2: ProdRequire * Prod[sku:=sku2]));

  Capacity[time,resource]: SUM(sku: ResourceReq * Prod) <= 1;

END
```

7.3.3 SCPc

To build up the **SCPc** model as shown in Figure 5.1 several things are needed that are similar to the artificial "+1" that was inserted into the objective function for **mrp**. To avoid small production quantities at the last time buckets we add a new "Sunset" constraint. This constraint ensures there will be no production quantities for periods that would result in completion after the time horizon consisting of T time buckets. Such production quantities do not incur any cost penalty in the objective function, so although they also do no good, the solver might set some quantities during the search for an optimal solution. To avoid this nuisance, we add a constraint that would be

$$\sum_{\tau=T-LT(i)+1}^{T} x_{i,\tau} = 0 \qquad i = 1, \ldots, P$$

in the **SCPc** model.

```
Sunset[sku]: SUM(time: Prod
             WHERE (time >= Numbertime-Leadtime+1)) = 0;
```

Because the inventory after the last time bucket does not result in any costs we add another constraint to bring the inventory to normal levels at the end of the planning horizon as discussed on page 65. The constraint makes use of the data element IT, which stands for inventory tolerance:

```
     InitialInventory * (1 - Invtolerance)
  <=
     Invplus[sku,time:=last(time)]
  <=
     InitialInventory * (1 + Invtolerance);
```

Due to the nature of the constraints, it is useful to have a non-zero (although perhaps very small) value for all $C(i)$. The reason is that the material requirements constraints and the inventory constraints are "one-sided" in the sense that production of components is required for assembly and production is required in order to create inventory, but there is nothing to prevent spurious non-zero production quantities that are not subsequently in inventory or used as components. This can be repaired by adding constraints, but we think a better thing to do is to continue in the spirit of inserting "+1" into the objective function in **mrp**. We simply make sure that there is a small penalty for all production, so that any production that appears in the plan is there for a reason and is not simply an artifact of the computer's solution process.

Notice that rather than using the inventory macro, we have expanded it. For example, the mrp constraint uses the expression for the macro as given on page 48.

The MPL implementation of **SCPc** is as follows:

```
TITLE
  SCPc;

INDEX
  sku := (AJ8172, LQ8811, RN0098, NN1100, WN7342);
  sku2 := sku;
  time := (jan1,jan2,jan3,jan4,jan5,jan6,jan7,jan8);
  resource := (HR_101,MT_402);

DATA
  Numbertime := count(time);

  InitialInventory[sku] := (90, 300, 100, 0, 900);

  Leadtime[sku]         := (2, 3, 4, 1, 12);

  ProdRequire[sku, sku2] := (0, 0, 0, 0, 0, ! R(sku, sku2)
                             2, 0, 0, 0, 0,
                             1, 0, 0, 0, 0,
                             0, 1, 0, 0, 0,
                             0, 1, 0, 0, 0);

  ExtDemand[sku,time] := (20, 30, 10, 20, 30, 20, 30, 40,
                          0,  0,  0,  0,  0,  0,  0,  0,
                          0,  0,  0,  0,  0,  0,  0,  0,
                          0,  0,  0,  0,  0,  0,  0,  0,
                          0,  0,  0,  0,  0,  0,  0,  0);

  ResourceReq[sku,resource] := (0.00208333, 0,
                                0.00104166, 0.00333333,
                                0,          0,
                                0.000001,   0,
                                0,          0);
```

```
  ordinal[time] := (1,2,3,4,5,6,7,8);
  LM := 10000;
  Invtolerance := 0.25;
  MaxExtraResource := 0.5; !no need for vector as is scalar value
  ChangeoverWaste[sku,sku2] := (0, 0, 0, 0, 0,
                                10,10,0, 0, 0,
                                0, 0, 0, 0, 0,
                                0, 0, 0, 0, 0,
                                0, 0, 0, 0, 0);
  ResChangeUsage[sku,resource] := (0.2, 0,
                                   0.2, 0.5,
                                   0,   0,
                                   0,   0,
                                   0,   0);

  HoldingCost[sku] := (2, 1, 0.5, 0.1, 0.1);
  ChangeoverCost[sku] := (800, 4200, 1, 1, 1);
  AddFractionCost := 1;
  TardinessCost[sku] := (400, 100, 4, 4, 4);

DECISION VARIABLES
  Prod[sku,time];
  PrdIndicator[sku,time];
  Invminus[sku,time];
  Invplus[sku,time];
  Overtime[resource,time];


MODEL
  MIN Cost = SUM(time: SUM(sku: (TardinessCost * Invminus +
                                 HoldingCost * Invplus +
                           ChangeoverCost * PrdIndicator) +
             SUM(resource: AddFractionCost * Overtime)));

SUBJECT TO
  MaterialRequirement[sku,time]:
       SUM(time=1..time-Leadtime: Prod)
     + InitialInventory
     + SUM(time=1..time: ExtDemand)
    >=
       SUM(time=1..time: ExtDemand
         + SUM(sku2: ProdRequire * Prod[sku:=sku2]
              + ChangeoverWaste * PrdIndicator[sku:=sku2]));

  Sunset[sku]: SUM(time: Prod
               WHERE (time >= Numbertime-Leadtime+1)) = 0;

  Capacity[time,resource]:
       SUM(sku: ResourceReq * Prod + ResChangeUsage * PrdIndicator)
                                                   - Overtime <= 1;

  MaxFraction[resource,time]:
```

```
      Overtime - MaxExtraResource <= 0;

  ProductionIndicator[sku,time]:
      PrdIndicator - (Prod / LM) >= 0;

  InventoryCalc[sku,time]:
      Invplus - Invminus -
     (SUM(time=1..time-Leadtime: Prod)
    + InitialInventory
    - SUM(time=1..time: ExtDemand
        + SUM(sku2: ProdRequire * Prod[sku:=sku2]
        + ChangeoverWaste * PrdIndicator[sku:=sku2]))) = 0;

  InventoryTolerance[sku]:
     InitialInventory * (1 - Invtolerance)
  <=
     Invplus[sku,time:=last(time)]
  <=
     InitialInventory * (1 + Invtolerance);

BINARY
  PrdIndicator;

END
```

7.4 OPL

The OPL modeling language is part of the modeling and solver environment supported by ILOG. To obtain more information concerning OPL and ILOG see `http://oplstudio.ilog.com`.

OPL statements generally end with a semicolon. Comments are delimited as in C++; that is, any text between /* and */ is a comment as is any text on a line after //. Comments may appear anywhere in a file. The definitions used by our models are grouped using the following sections:

- Data declaration — Data are declared to have a type and, for vectors, indexes of an appropriate type. These statements begin with a keyword for a simple type or else a set bracket enclosed structured type. For example, the statement

  ```
  {string} PP = ...;        // Set of SKUs
  ```

  declares a list of SKUs. The use of an ellipsis on the right hand side indicates that the values will be supplied later from some other source such as a file or database. The statement

  ```
  int D[PP,TT] = ...;       // External demand for SKU i in period t
  ```

  declares the existence of a vector of integers indexed by `TT` and `PP` whose values will be supplied later. These straightforward declarations are easy to understand. However, when the tables are largely filled with zeros, such as one would expect for R and often D, then a *sparse* representation makes more sense than the *dense* representation shown here. The authors' web site (see page 81) has examples of both. In the sections that follow we also provide examples of each.
- Decision variables — Decision variables are declared in statements that begin with the keyword "var" followed by a type and then the variable name and a declaration of its indexes. So

  ```
  var float+ x[PP,TT];        // Order quantity for SKU i, time t
  var int d[PP,TT] in 0..1;   // Prod. indicator for SKU i, time t
  ```

  declare `x` to be indexed by parts and periods and to take on values that are positive floating point numbers (i.e., not necessarily integers, but not negative) and `d` to be binary with the same index sets.
- Model declaration — A keyword such as `minimize` signals the start of the statement of the objective function. So, for example,

  ```
  minimize
      sum (i in PP, t in TT) (T-t)*x[i,t]
  ```

  declares the **mrp** objective function. The keywords `subject to` mark the constraints. OPL supports statement blocks delimited by set brackets, so the line

```
subject to {
```

begins the constraints and

```
forall (i in PP, t in TT) {
```

begins the declaration of those constraints that apply to all products and periods (which, for **mrp**, happens to be all of the constraints). Each block that is opened with a left set bracket is closed with a right one. The constraints themselves are declared in a straightforward way; for example

```
forall (i in PP, t in TT) {
    x[i,t] - d[i,t]*LS[i] >= 0;
}
```

declares

$$x_{i,t} - \delta_{i,t} LS(i) \geq 0 \;\; i = 1, \ldots, P, \;\; t = 1, \ldots, T.$$

7.4.1 mrp

Our OPL implementation makes use of three files. The first contains the model, the second contains the data and the third contains project information, the most important of which is the location of the first two files.

Model file. A few comments about the implementation of the **mrp** model may be helpful:

1. To avoid small production quantities in the last time buckets we insert "+1" into the objective function, which insures that production is always penalized a little bit. For additional discussion, see page 87.
2. The lack of a semicolon after the objective function and the presence of one after the last set bracket might seem strange at first glance; however, the syntax is sensible from a technical perspective. It turns out that `minimize subject to` is a single statement that can contain blocks of statements within.

```
/*** OPL Implementation of mrp ***/


/*** DATA DECLARATION ***/
{string} PP = ...;        // Set of SKUs
int M = ...;              // Large constant used inside constraints
int T = ...;              // Number of time buckets
range TT 1..T;            // Range of time periods

int I[PP] = ...;          // Initial inventory for SKU i
int LT[PP] = ...;         // Lead time for SKU i
int R[PP,PP] = ...;       // Amount of SKU i needed to make one j
int D[PP,TT] = ...;       // External demand for SKU i in period t
int LS[PP] = ...;         // Lot size for SKU i


/*** DECISION VARIABLES ***/
var float+ x[PP,TT];        // Order quantity for SKU i, time t
var int d[PP,TT] in 0..1;   // Production indicator for SKU i, time t


/*** OPTIMIZATION MODEL ***/

minimize
    sum (i in PP, t in TT) (T-t+1)*x[i,t]

subject to {

    forall (i in PP, t in TT) {
       // Demand and materials requirement
       sum (r in 1..t-LT[i])
           x[i,r]+ I[i]
           - sum (r in 1..t) (D[i,r] + sum (j in PP) R[i,j]*x[j,r])
           >= 0;

        // Lot size requirement
        x[i,t] - d[i,t]*LS[i] >= 0;

        // Modeling constraint for production indicator
        M*d[i,t] - x[i,t] >= 0;
    }
};
```

Data file. The data file shown here provides values for the data elements defined in the **mrp** model file. The syntax for R and D makes sense when one thinks of two dimensional data structures as vectors of vectors.

```
/* OPL data for mrp model */

PP = {"AJ8172", "LQ8811", "RN0098", "NN1100", "WN7342"}; // SKUs
T = 8;                              // Time buckets
M = 10000;                          // Large constant

I = [90,300,100,0,900];             // Initial inventory
LT = [2,3,4,1,12];                  // Lead time for SKU i
LS = [100,400,100,1,1000];          // Lot size for SKU i

// Amount of SKU i needed to make one j
R = [
   [0,0,0,0,0],
   [2,0,0,0,0],
   [1,0,0,0,0],
   [0,1,0,0,0],
   [0,1,0,0,0]
];

// External Demand for SKUs in each period
D = [
   [20,30,10,20,30,20,30,40],
   [0,0,0,0,0,0,0,0],
   [0,0,0,0,0,0,0,0],
   [0,0,0,0,0,0,0,0],
   [0,0,0,0,0,0,0,0]
];
```

Project file. In the interest of completeness, we provide the OPL project file for this example. Normally, project files are created by the graphical user interface and are not edited directly. The details of the project file syntax are not important for our purposes.

```
defaultDirectory = "D:\My Documents\My Work\OPL Studio\"
display_activities = 0
display_result = 0
display_sorted_activities = 0
dta_font = "StockedSystemFixed"
edt_font = "StockedSystemFixed"
mru0 = "D:\My Documents\My Work\OPL Studio\mrp.dat"
mru1 = "D:\My Documents\My Work\OPL Studio\mrp.mod"
output_font = "StockedSystemFixed"
shw_font = "StockedDefaultGui"
MODEL = "mrp.mod"
DATA = "mrp.dat"
```

7.4.2 MRPII

The OPL implementation of **MRPII** is a straightforward extension of **mrp**, so we omit the data and project files, both of which are available through the authors' web site (see page 81).

```
/*** DATA DECLARATION ***/
{string} PP = ...;          // Set of SKUs
{string} KK = ...;          // Set of resources
int T = ...;                // Number of time buckets
range TT 1..T;              // Range of time periods

int I[PP] = ...;            // Initial inventory for SKU i
int LT[PP] = ...;           // Lead time for SKU i
int R[PP,PP] = ...;         // Amount of SKU i needed to make one j
int D[PP,TT] = ...;         // External demand for SKU i in period t

float U[PP,KK] = ...;       // Fraction of res. k for one unit of SKU i


/*** DECISION VARIABLES ***/
var float+ x[PP,TT];        // Order release for SKU i in time t


/*** OPTIMIZATION MODEL ***/

minimize
    sum (i in PP, t in TT) (T-t+1)*x[i,t]

subject to {

   // Requirement constraint
    forall (t in TT, i in PP)
        sum (r in 1..t-LT[i]) x[i,r]+ I[i]
          - sum (r in 1..t) (D[i,r] + sum (j in PP) R[i,j]*x[j,r])
         >= 0;

    // Capacity constraint
    forall (k in KK, t in TT)
        sum (i in PP) (U[i,k] * x[i,t]) <= 1;
};
```

7.4.3 SCPc

The **SCPc** model as shown in Figure 5.1 is rich enough to afford many options for implementation. The authors' web site (see page 81) contains a total of four OPL implementations of **SCPc**: There are sparse and direct translations and for each of these there is a version that uses two inventory variables as in the model shown here and one that uses OPL's piecewise linear objective function so that there is no need to divide the inventory variable into a positive and negative component.

Model for SCPc. Here we show the sparse version with the piecewise linear objective function. A number of points are helpful in understanding this implementation:

1. The sparse version relies on structures rather than simple vectors. The structures enable the association of names with data in a list that contains only entries for data that are present. For example, the statements

```
struct rStr { string skuFrom; string skuTo; int qty; };
{rStr} RR = ...;    // Amount of SKU i needed to make one j
```

   create a table called RR to provide a sparse representation of R.
2. The ability to specify a piecewise-linear objective function obviates the need for the two inventory variables, so this formulation is slightly different than the one given earlier for **SCPc**. The piecewise term is given by

```
piecewise{ -A[i] -> 0; H[i] } inv[i,t]
```

   which means: use `-A[i]` times `inv[i,t]` when `inv` is negative (the symbol "->" means "up to") and `H[i]` times `inv[i,t]` when `inv` is positive.

```
/*** DATA DECLARATION ***/
{string} PP = ...;      // Set of SKUs
{string} KK = ...;      // Set of resources
int M = ...;            // Large constant used inside constraints
int T = ...;            // Number of time buckets
range TT 1..T;          // Range of time periods
float IT = ...;         // Tolerance on final inventory

int I[PP] = ...;        // Initial inventory for SKU i
int LT[PP] = ...;       // Lead time for SKU i

float F[KK,TT] = ...; // Max. frac. of res. k that can be added

int O[KK,TT] = ...;     // Marginal cost per cap. frac. added to k
float H[PP] = ...;      // Per period holding cost for SKU i
int C[PP] = ...;        // Total (out of pocket) changeover cost
int A[PP] = ...;        // Per period tardiness cost for SKU i

struct rStr { string skuFrom; string skuTo; int qty; };
struct dStr { string sku; int period; int qty; };
struct uStr { string sku; string res; float qty; };
struct sStr { string sku; string res; float qty; };
struct wStr { string skuFrom; string skuTo; int qty; };
{rStr} RR = ...;        // Amount of SKU i needed to make one j
{dStr} DD = ...;        // External demand for SKU i in period t
{uStr} UU = ...;        // Frac. of res. k needed to make one i
{sStr} SS = ...;        // Frac. of res. k to change to SKU i
{wStr} WW = ...;        // Waste of SKU i to change to SKU j
```

```
/*** DECISION VARIABLES ***/
var float+ x[PP,TT];       // Release quantity for SKU i in time t
var float+ y[KK,TT];       // "Overtime" fraction of res. k in t
var int d[PP,TT] in 0..1;  // Indicator of prod. of SKU i in t
var float inv[PP,TT];      // Inv. of i at t; backlog is negative


/*** OPTIMIZATION MODEL ***/

// Use piecewise linear term for inventory cost:
// H[i] when inv[i,t] >= 0 and -A[i] when inv[i,t] < 0

minimize
    sum (t in TT, i in PP) (piecewise{ -A[i] -> 0; H[i] } inv[i,t]
                             + C[i]*d[i,t]) +
    sum (t in TT, k in KK) O[k,t]*y[k,t]

subject to {

    forall (t in TT, i in PP) {

        // MRP constraint
        inv[i,t] + sum (<i,r,dd> in DD : r <= t) dd >= 0;

        // Production indicator
        M*d[i,t] - x[i,t] >= 0;

        // Inventory constraints
        inv[i,t] =
            sum (r in 1..t-LT[i]) x[i,r]
                  + I[i]
                  - sum (<i,r,dd> in DD : r <= t) dd
                  - sum (r in 1..t, <i,j,rr> in RR) rr*x[j,r]
                  - sum (r in 1..t, <i,j,ww> in WW) ww*d[j,r];
    };

    // Capacity limits
    forall (t in TT, k in KK) {
        sum (<i,k,uu> in UU) uu*x[i,t]
        + sum (<i,k,ss> in SS) ss*d[i,t] <= 1 + y[k,t];

        // Overtime limit
        y[k,t] <= F[k,t];
    };

    forall (i in PP) {
        // avoid "Random" production for ending conditions
        sum (s in [T-LT[i]+1..T]) x[i,s] = 0;

        // Meet final inventory targets
        (1-IT)*I[i] <= inv[i,T] <= (1+IT)*I[i];
    };
};
```

Data for SCPc. The **SCPc** data file provides values for the structures defined in the model file. For example,

```
RR = {
    <"LQ8811", "AJ8172", 2>,
    <"RN0098", "AJ8172", 1>,
    <"NN1100", "LQ8811", 1>,
    <"WN7342", "LQ8811", 1>
};
```

establishes the value for the requirements, R.

```
/* OPL data for SCPc model */

PP = {"AJ8172", "LQ8811", "RN0098", "NN1100", "WN7342"}; // SKUs
KK = {"HR-101", "MT-402"};        // Resources
T = 8;                            // Time buckets
M = 10000;                        // Large constant
IT = 0.25;                        // Tolerance on final inventory

I = [90,300,100,0,900];           // Initial inventory
LT = [2,3,4,1,12];                // Lead time for SKU i

// Amount of SKU i needed to make one j
RR = {
    <"LQ8811", "AJ8172", 2>,
    <"RN0098", "AJ8172", 1>,
    <"NN1100", "LQ8811", 1>,
    <"WN7342", "LQ8811", 1>
};

// External Demand for SKUs in each period
DD = {
    <"AJ8172",1,20>,
    <"AJ8172",2,30>,
    <"AJ8172",3,10>,
    <"AJ8172",4,20>,
    <"AJ8172",5,30>,
    <"AJ8172",6,20>,
    <"AJ8172",7,30>,
    <"AJ8172",8,40>
};

// Fraction of resources needed to make each SKU
UU = {
    <"AJ8172", "HR-101", 0.00208333>,
    <"LQ8811", "HR-101", 0.00104166>,
    <"LQ8811", "MT-402", 0.00333333>,
    <"NN1100", "HR-101", 0.000001>
};
```

```
// Maximum fraction of a resource that can be added in each period
F = [
   [.5, .5, .5, .5, .5, .5, .5, .5],
   [.5, .5, .5, .5, .5, .5, .5, .5]
];

// Fraction of resource required to changeover to SKU
SS = {
     <"LQ8811", "MT-402", 0.5>,
     <"AJ8172", "HR-101", 0.2>,
     <"LQ8811", "HR-101", 0.2>
};

// Waste when changing between SKUs
WW = {
    <"LQ8811", "AJ8172", 10>,
    <"LQ8811", "LQ8811", 10>
};

// Marginal cost per capacity fraction added to resource k
O = [
   [1, 1, 1, 1, 1, 1, 1, 1],
   [1, 1, 1, 1, 1, 1, 1, 1]
];

H = [2, 1, 0.5, 0.1, 0.1];       // Per period holding costs
C = [800, 4200, 1, 1, 1];        // Changeover costs
A = [400, 100, 4, 4, 4];         // Tardiness costs
```

7.5 Xpress-Mosel

Dash Optimization's Xpress-Mosel language provides algebraic modeling and the features of a programming language. Mosel is based on an open, modular architecture that makes it possible to add access to solvers of various types, databases, or any other specific functionality (e.g., system functions, new data types) to the language in the form of *modules.* A graphical interface, Xpress-IVE, is available for working with Mosel models. See `http://www.dashoptimization.com` for more information about Mosel, the available modules, and the Xpress-MP product family.

A Mosel model starts with the keyword `model` followed by the name of the model, and terminates with `end-model`. The definition of a model consists of a sequence of model statements such as constraint definition, solver commands, and blocks (e.g., `declarations`, `initializations`, and subroutines). Any text following on the same line after the character `!` is a comment that is ignored by the Mosel compiler. Multi-line comments are surrounded by `(!` and `!)`.

Conceptually, a Mosel model file can be divided into the following sections:

- Data — In the example problems the data sets are small so the data arrays could be assigned directly in the model. However, to obtain easily reusable models that can be run with different data sets it is preferable to read in all data from a separate file. The data arrays and the corresponding index sets need to be declared in a `declarations` block. If the contents of the index sets is not known at the creation of an array, the latter is *dynamic.* This is typically the case when the array and its index sets are initialized from file as in the following code extract

```
declarations
 PP: set of string
 I: array(PP) of real
end-declarations

initializations from 'mrp.dat'
 I
end-initializations
```

  An advantage of this definition is that only the array entries listed in the data file will be created. This is referred to as *sparse* format. General sets are unordered (i.e., it is not possible to determine something like a "last" element). The set type `range` defines an ordered set of integers:

```
 TT: range
```

  In the model examples in this section, range sets are used for the time periods (indices $1, ..., T$ as in the mathematical models), while the sets of SKUs and resources do not require any ordering relation and contain, therefore, simply the names of the products.

- Variables — Mosel variables are declared in a fashion similar to data array declarations, using the type `mpvar`. To define the variables $d_{i,t}$ with i in PP and t in TT we write:

```
declarations
 d: array(PP,TT) of mpvar
end-declarations
```

The index sets should preferably be known at the declaration of the variables; we therefore declare the variables after reading the data from file. To define variables $d_{i,t}$ as binaries, we add the following line after their declaration:

```
forall(i in PP,t in TT) d(i,t) is_binary
```

- Constraints — A typical linear constraint like

$$\sum_{i=1}^{P}\sum_{k=1}^{K} U_{ik} \cdot x_{it} \leq 1 \quad t = 1, ..., T$$

is written in Mosel as

```
forall(t in 1..T) sum(i in 1..P,k in 1..K) U(i,k)*x(i,t) <= 1
```

In the formulation of linear constraints the operator signs `<=`, `>=`, and `=` may be used. Variables may appear on either side of the operator sign. Optionally, constraints may be named. This is especially useful if one wishes to refer to them later on in the model, for instance, to modify their definition. Unnamed constraints on a single variable are interpreted by Mosel as bounds which are commonly treated by LP/MIP solvers in a different, more efficient way than linear constraints. The objective function usually is defined as a constraint without operator and right hand side.

- Solver Commands — In the examples, we use Xpress-Optimizer for solving the problems. The corresponding Mosel module needs to be loaded at the beginning of every model:

```
uses "mmxprs"
```

The solver module *mmxprs* provides, among others, the procedures such as `minimize` to solve LP or MIP problems.

- Solution Output Commands — The procedures `write` and `writeln` may be used for displaying results. The solution value of a variable is obtained with `getsol`. The following prints the solution value of the variable $x_{i,t}$:

```
writeln(getsol(x(i,t)))
```

In the examples, the function `strfmt` is used to format the printed output by indicating the variable name and the number of columns to be used.

- Subroutines — To structure larger models, especially if they involve some calculations for the data or (parts of) solution algorithms implemented in the Mosel language, Mosel lets the user define two types of subroutines: procedures (no return value) and functions. Subroutines may take parameters, their structure is similar to the Mosel model itself.
- Loops and selections — Besides the simple `forall` loop that is frequently used for stating constraints, Mosel also defines a multi-line version `forall-do` of this loop, and `repeat-until` and `do-while` loops. The calculation of the big M values in the **mrp** model in the following subsection gives an example of using these loops. In the same model, also the use of the selection statement `if-then-else` is demonstrated.
  A common feature of algebraic modeling languages is the possibility to restrict a set of indices with logical conditions. In Mosel, the vertical bar `|` is used to indicate such conditions:

  ```
  sum(s,t in 1..T | s<=t)
  ```

  represents the double sum

  $$\sum_{t=1}^{T}\sum_{s=1}^{t}$$

  Logical conditions may be combined with `and` and `or`.

7.5.1 mrp Model

The following Mosel implementation of **mrp** defines the mathematical model, then solves it and outputs the result. As in the GAMS implementation, we calculate a specific big M value `M(i)` for every SKU i instead of using the same, very large value in all cases. The calculation of these values is separated from the model by placing it into a procedure.
A few notes on the implementation:

1. After reading in the data, the index sets are *finalized*. This means, the contents of these sets cannot change. The variables that are subsequently defined with these index sets are then handled in a more efficient way by Mosel.
2. The procedure `calculate_bigM` is used before it is defined. Therefore, we need to declare it at the beginning using the keyword `forward`.
3. The calculation of the SKU-specific big M values `M(i)` follows the same idea as the algorithm used in the GAMS implementation. We first determine a bound on the total demand of an SKU:
   - For a root node `i` in the production tree (a product not used in the production of any other) an upper bound on the total demand `TD(i)` is given by the sum of the external demand `D(i,t)` over all time periods plus the lot size. These nodes are collected in the set `CurLevel`. All other nodes are put into the set `TreeNodes`. Note that

a production tree refers to a BOM with a special structure (e.g., a convergent structure).

– For all nodes `i` in `CurLevel`: go through the remaining nodes in `TreeNodes` to find those nodes `j` that are used for the production of a node `i` in `CurLevel`. The demand for such a node `j` is given by its external demand plus the implicit demand through the production of `i`. The nodes `j` are collected into the set `NextLevel`.

  Make `NextLevel` the new set `CurLevel` and remove these nodes from `TreeNodes`.

  Repeat the loop until all nodes have been enumerated (i.e., the set `CurLevel` becomes empty).

For every `i`, the value of `M(i)` results from the upper bound on total demand, `TD(i)`.

4. The objective function is defined using

```
Obj:= sum(i in PP,t in TT) (getsize(TT) - t + 1) * x(i,t)
```

To avoid small production quantities in the last time buckets we insert "+1" into the objective function, which insures that production is always penalized a little bit. For additional discussion, see page 87. The function `getsize(TT)` returns the number of elements in the range set `TT` and hence the number of time buckets.

```
model "mrp"
 uses "mmxprs"                        ! Use Xpress-Optimizer for solving

 forward procedure calculate_bigM

 declarations
  TT: range                           ! Time Buckets
  PP: set of string                   ! SKUs

  M: array(PP) of real                ! Big M
  R: array(PP,PP) of real             ! Number of SKUs i to produce a j
  D: array(PP,TT) of real             ! External demand
  I: array(PP) of real                ! Beginning inventory of SKU i
  LS: array(PP) of real               ! Lot size of item i
  LT: array(PP) of real               ! Lead time of item i
 end-declarations

 initializations from 'mrp.dat'
  R D I LS LT
 end-initializations

 finalize(TT); finalize(PP)           ! Finalize index sets

 calculate_bigM                       ! Calculate values for M

 declarations
  d: array(PP,TT) of mpvar            ! Production indicator
```

```
 x: array(PP,TT) of mpvar      ! Number of SKUs to produce
end-declarations

Obj:= sum(i in PP,t in TT) (getsize(TT) - t + 1) * x(i,t)

forall(i in PP,t in TT)
 MaterialRequirement(i,t):=
  sum(s in TT | s <= t - LT(i)) x(i,s) + I(i) >=
   sum(s in TT | s <= t) ( D(i,s) + sum(j in PP) R(i,j)*x(j,s) )

forall(i in PP,t in TT)
 LotSize(i,t):= x(i,t) >= d(i,t)*LS(i)

forall(i in PP,t in TT)
 ProductionIndicator(i,t):= d(i,t) >= x(i,t)/M(i)

! Integrality constraint for production indicator
 forall(i in PP,t in TT) d(i,t) is_binary

! Solve the problem
 minimize(Obj)

! Solution printing
 writeln("Objective value: ", getobjval)
 writeln("Production plan:")
 write("Period")
 forall(t in TT) write(strfmt(t,5))
 writeln
 forall(i in PP) do
  write(i)
  forall(t in TT) write(strfmt(getsol(x(i,t)),5))
  writeln
 end-do

!-------------------------------------------------------------

! Calculate values for M
 procedure calculate_bigM
  declarations
   CurLevel: set of string       ! Nodes in current tree level
   NextLevel: set of string      ! Nodes in next deeper tree level
   TreeNodes: set of string      ! Remaining tree nodes
   TD: array(PP) of real         ! Upper bound on total demand
  end-declarations

  forall(i in PP)
   if sum(j in PP) R(i,j)>0 then
    TreeNodes+={i}
   else
    TD(i):=sum(t in TT) D(i,t) + LS(i)
                                 ! Total demand for root node(s)
    CurLevel+={i}
   end-if
```

```
    repeat
     forall(i in CurLevel) do
      NextLevel:={}
      forall(j in TreeNodes | exists(R(j,i))) do
       NextLevel+={j}
       TD(j):= sum(t in TT) D(j,t) + R(j,i)*TD(i) + LS(i)
                                              ! Sum for nodes
      end-do
     end-do
     CurLevel:=NextLevel
     TreeNodes-=NextLevel
    until (CurLevel={})

    forall(i in PP) M(i) := TD(i)    ! Set M

  end-procedure

end-model
```

7.5.2 mrp Data

The following data file `mrp.dat` is used by the Mosel implementation of the **mrp** model and also by the **MRPII** model in the following section. Every data array is labeled by its name. Since all index sets are initialized dynamically from the data file we need to indicate the index-tuple for every entry. Only the non-zero entries need to be given in the data file (sparse format). As in the Mosel model files, the exclamation sign `!` marks comments.

```
! External demand
D: [ (AJ8172 1) 20 (AJ8172 2) 30 (AJ8172 3) 10 (AJ8172 4) 20
     (AJ8172 5) 30 (AJ8172 6) 20 (AJ8172 7) 30 (AJ8172 8) 40]

! Lead time of items
LT: [ (AJ8172) 2 (LQ8811) 3 (RN0098) 4 (NN1100) 1 (WN7342) 12 ]

! Lot sizes
LS: [ (AJ8172) 100 (LQ8811) 400 (RN0098) 100
      (NN1100) 1 (WN7342) 1000 ]

! Beginning inventory of SKUs
I: [ (AJ8172) 90 (LQ8811) 300 (RN0098) 100 (WN7342) 900 ]

! Number of i to produce one j
R: [ (LQ8811 AJ8172) 2
     (RN0098 AJ8172) 1
     (NN1100 LQ8811) 1
     (WN7342 LQ8811) 1 ]

! Fraction of resources needed by SKUs (MRPII model)
U: [(AJ8172 "HR-101") 0.00208333
    (LQ8811 "HR-101") 0.00104166
```

```
(LQ8811 "MT-402") 0.00333333
(NN1100 "HR-101") 0.00000100 ]
```

7.5.3 mrp Results

The model and data file together produce the following output:

```
Objective value: 6800
Production plan:
Period     1    2    3    4    5    6    7    8
AJ8172     0    0  100    0    0  100    0    0
LQ8811     0    0  400    0    0    0    0    0
RN0098     0  100    0    0    0    0    0    0
NN1100     0  400    0    0    0    0    0    0
WN7342     0    0    0    0    0    0    0    0
```

7.5.4 MRPII Model

The following Mosel program implements the model **MRPII**. It is possible to formulate and solve the two models **mrp** and **MRPII** in a single Mosel program, similar to the GAMS implementation. However, to keep things simple, the **MRPII** model is given as a separate program, although in this implementation they use the same data file.

```
model "MRP2"
 uses "mmxprs"                        ! Use Xpress-Optimizer for solving

 declarations
  TT: range                           ! Time Buckets
  PP: set of string                   ! SKUs
  KK: set of string                   ! Resources

  R: array(PP,PP) of real             ! Number of SKUs i for one SKU j
  D: array(PP,TT) of real             ! External demand
  I: array(PP) of real                ! Beginning inventory of SKU i
  U: array(PP,KK) of real             ! Frac. of res. k for one SKU i
  LT: array(PP) of real               ! Lead Time of item i
 end-declarations

 initializations from 'mrp.dat'
  R D I U LT
 end-initializations

 finalize(TT); finalize(PP)           ! Finalize index sets

 declarations
  x: array(PP,TT) of mpvar            ! Number of SKUs to produce
 end-declarations

 Obj:= sum(i in PP,t in TT) (getsize(TT) - t + 1) * x(i,t)
```

```
 forall(i in PP,t in TT)
  MaterialRequirement(i,t):=
   sum(s in TT | s <= t - LT(i)) x(i,s) + I(i) >=
    sum(s in TT | s <= t) (D(i,s) + sum(j in PP) R(i,j)*x(j,s))

 forall(t in TT,k in KK)
  Capacity(t,k):=  sum(i in PP) U(i,k)*x(i,t) <= 1

! Solve the problem
 minimize(Obj)

! Solution printing
 writeln("Objective value: ", getobjval)
 writeln("Production plan:")
 write("Period")
 forall(t in TT) write(strfmt(t,5))
 writeln
 forall(i in PP) do
  write(i)
  forall(t in TT) write(strfmt(getsol(x(i,t)),5))
  writeln
 end-do

end-model
```

7.5.5 SCPc Model

To build up the **SCPc** model as shown in Figure 5.1 several things are needed that are similar to the artificial "+1" that was inserted into the objective function for **mrp**. To avoid small production quantities at the last time buckets we add a new "Sunset" constraint. This constraint ensures there will be no production quantities for periods that would result in completion after the time horizon consisting of T time buckets. Such production quantities do not incur any cost penalty in the objective function, so although they also do no good, the solver might set some quantities during the search for an optimal solution. To avoid this nuisance, we add a constraint that would be

$$\sum_{\tau=T-LT(i)+1}^{T} x_{i,\tau} = 0 \qquad i = 1, \ldots, P$$

in the **SCPc** model.

```
forall(i in PP)
 Sunset(i):=
  sum(s in TT | s > getsize(TT) - LT(i) + 1) x(i,s) = 0
```

Because the inventory after the last time bucket does not result in any costs we add another constraint to bring the inventory to normal levels at the end of the planning horizon as discussed on page 65. The constraint makes use of the data element IT, which stands for inventory tolerance:

```
forall(i in PP) do
  InventoryTol1(i):= (1-IT)*I(i) <= iplus(i,getlast(TT))
  InventoryTol2(i):= iplus(i,getlast(TT)) <= (1+IT)*I(i)
 end-do
```

Due to the nature of the constraints, it is useful to have a non-zero (although perhaps very small) value for all $C(i)$. The reason is that the material requirements constraints and the inventory constraints are "one-sided" in the sense that production of components is required for assembly and production is required in order to create inventory, but there is nothing to prevent spurious non-zero production quantities that are not subsequently in inventory or used as components. This can be repaired by adding constraints, but we think a better thing to do is to continue in the spirit of inserting "+1" into the objective function in **mrp**. We simply make sure that there is a small penalty for all production, so that any production that appears in the plan is there for a reason and is not simply an artifact of the computer's solution process.

The following Mosel program implements the **SCPc** model. Contrary to our implementation of the **mrp** model, we use a single big M value, the value of which is read from the data file. Instead of the inventory macro, an additional variable is used in this implementation.

```
model "SCPc"
 uses "mmxprs"                       ! Use Xpress-Optimizer for solving

 declarations
  TT: range                          ! Time Buckets
  PP: set of string                  ! SKUs
  KK: set of string                  ! Resources

  M: integer                         ! Large number
  IT: real                           ! Inventory tolerance at the end
  R: array(PP,PP) of real            ! SKUs i to produce one j
  D: array(PP,TT) of real            ! Demand for an item in a period
  I: array(PP) of real               ! Beginning inventory of SKU i
  U: array(PP,KK) of real            ! Frac. of res. k for one SKU i
  LT: array(PP) of real              ! Lead time of item i
  F: array (KK,TT) of real           ! Max. frac. of k addable in t
  S: array (PP,KK) of real           ! Frac. of k to change to SKU i
  W: array (PP,PP) of real           ! Waste of i to change to j
  H: array (PP) of real              ! Period holding cost for SKU i
  C: array (PP) of real              ! Total changeover cost for SKU i
  O: array (KK,TT) of real           ! Marginal cost per fraction added
  A: array (PP) of real              ! Period tardiness cost
 end-declarations

 initializations from 'scp.dat'
  M IT R D I U LT F S W H C O A
 end-initializations

 finalize(TT); finalize(PP); finalize(KK)   ! Finalize index sets
```

```
 declarations
  d: array(PP,TT) of mpvar       ! Production indicator
  x: array(PP,TT) of mpvar       ! Number of SKUs to produce
  y: array(KK,TT) of mpvar       ! "Overtime" frac. of ress k in t
  iplus: array(PP,TT) of mpvar   ! Inventory of SKU i to carry in t
  iminus: array(PP,TT) of mpvar  ! Quantity of i backordered in t
  inv: array(PP,TT) of mpvar     ! Inventory variables
 end-declarations

! Objective function
 Cost:= sum(t in TT) (sum(i in PP) (A(i)*iminus(i,t) +
                                    H(i)*iplus(i,t) +
                                    C(i)*d(i,t)) +
                      sum(k in KK) O(k,t)*y(k,t))

 forall(i in PP,t in TT)
  DefInventory(i,t):=
   inv(i,t) = sum(s in TT | s <= t - LT(i)) x(i,s) + I(i) -
    sum(s in TT | s <= t) (D(i,s) + sum(j in PP) (R(i,j) * x(j,s) +
                                                  W(i,j) * d(j,s)) )

 forall(i in PP,t in TT)
  MaterialRequirement(i,t):=
   inv(i,t) >= - sum(s in TT | s <= t) D(i,s)

 forall(t in TT,k in KK)
  Capacity(t,k):=
   sum(i in PP) (U(i,k)* x(i,t) + S(i,k)*d(i,t)) <= 1 + y(k,t)

 forall(k in KK,t in TT)
  MaxFraction(k,t):= y(k,t) <= F(k,t)

 forall(i in PP,t in TT)
  ProductionIndicator(i,t):= d(i,t) >= x(i,t)/M

 forall(i in PP,t in TT)
  Inventory(i,t):= iplus(i,t) - iminus(i,t) = inv(i,t)

 forall(i in PP)
  Sunset(i):=
   sum(s in TT | s > getsize(TT) - LT(i) + 1) x(i,s) = 0

 forall(i in PP) do
  InventoryTol1(i):= (1-IT)*I(i) <= iplus(i,getlast(TT))
  InventoryTol2(i):= iplus(i,getlast(TT)) <= (1+IT)*I(i)
 end-do

! Inventory may take negative values
 forall(i in PP,t in TT) inv(i,t) is_free

! Integrality constraint for production indicator
 forall(i in PP,t in TT) d(i,t) is_binary

! Solve the problem
 minimize(Cost)
```

```
! Solution printing
 writeln("Total cost: ", getobjval)
 writeln("Production plan:")
 write("Period")
 forall(t in 0..getlast(TT)) write(strfmt(t,6))
 writeln("  Dev.")
 forall(i in PP) do
  write(i, "     -")
  forall(t in TT) write(strfmt(getsol(x(i,t)),6,1))
  write("\n  Inv.", strfmt(I(i),6,1))
  forall(t in TT) write(strfmt(getsol(inv(i,t)),6,1))
  writeln(strfmt(getsol(iminus(i,getlast(TT))),6,1))
 end-do

end-model
```

8. Solutions

We have now developed models with reasonable detail to be used for supply chain planning. Once we have a model, we need to get the data for it, find solutions to it, and perhaps provide information about the solutions to the software or people responsible for detailed planning and scheduling. The data must either come from ERP systems that store performance data or the data must be estimated by production and engineering staff. If a solution is to be implemented, then it must be provided to the ERP system or to production schedulers.

A modeling language can be used to put the model into a form suitable for solution by a computer. As mentioned earlier and as demonstrated in Chapter 7, these languages allow input of the model using notation that is very similar to the notation that we have used. Once the models are entered and the paths to the data are provided to the modeling languages, optimization software is called by the modeling language software to find solution values. For many problems, the commercial optimization software must be augmented or replaced by heuristic solution methods for reasons that will be clarified shortly.

Our main concern in this chapter is to outline methods for finding solutions. In order to create good models one does not need to know all of the details about solution techniques just as one does not need to understand all of the details of compiler technology in order to write computer software. However, in both cases it can be very helpful to know a little bit about what will happen to the model or program after it is written.

8.1 MIPs and Relaxations

All of the models that we have developed so far are linear models. We have used only models where variables are not multiplied by other variables and where powers and roots of variables are not used. Any useful model that has changeovers, minimum lot sizes, alternative routings or integer variables for any other reason is referred to as a mixed integer program (MIP). It is "mixed" because some of the variables must be integers, while others can take *real* values that are either fractions or integers.

To use in our discussions, we repeat the simple model **mrp** that was developed in §3.4.

$$\text{minimize:} \quad \sum_{i=1}^{P}\sum_{t=1}^{T}(T-t)x_{i,t} \qquad \textbf{(mrp)}$$

subject to:

$$\sum_{\tau=1}^{t-LT(i)} x_{i,\tau} + I(i,0) - \sum_{\tau=1}^{t}\left(D(i,\tau) + \sum_{j=1}^{P} R(i,j)x_{j,\tau}\right) \geq 0$$
$$i=1,\ldots,P,\ t=1,\ldots,T$$
$$x_{i,t} - \delta_{i,t}LS(i) \geq 0 \qquad i=1,\ldots,P,\ t=1,\ldots,T$$
$$\delta_{i,t} - \frac{x_{i,t}}{M} \geq 0 \qquad i=1,\ldots,P,\ t=1,\ldots,T$$
$$\delta_{i,t} \in \{0,1\} \qquad i=1,\ldots,P,\ t=1,\ldots,T$$
$$x_{i,t} \geq 0 \qquad i=1,\ldots,P,\ t=1,\ldots,T$$

This model is a MIP because it contains the variables δ each of which must take a value that is either zero or one. Such variables are often called *binary* variables. A MIP with only binary integer variables is called a *binary MIP*.

If we did not want to model lot sizes, the formulation would be as follows:

$$\text{minimize:} \quad \sum_{i=1}^{P}\sum_{t=1}^{T}(T-t)x_{i,t} \qquad \textbf{(lp mrp)}$$

subject to:

$$\sum_{\tau=1}^{t-LT(i)} x_{i,\tau} + I(i,0) - \sum_{\tau=1}^{t}\left(D(i,\tau) + \sum_{j=1}^{P} R(i,j)x_{j,\tau}\right) \geq 0$$
$$i=1,\ldots,P,\ t=1,\ldots,T$$
$$x_{i,t} \geq 0 \qquad i=1,\ldots,P,\ t=1,\ldots,T$$

This model has no integer variables and it is linear. Such models are often called *linear programs* or LPs. Note that LP is also used to refer to linear programming. The use of the word "program" is very unfortunate since it causes confusion between linear programs and computer programs. The word is used because the early models were developed to find schedules and plans and these are (or were) often referred to as programs.

Linear programming models are much easier to solve than MIPs. If one happens to be able to develop a supply chain production planning or scheduling model without integer variables, then powerful software is available that can solve very large instances of the model.

However, it is hard to imagine a supply chain production planning problem with no integers, so we focus our attention on MIPs. It happens that one typically solves linear programs as part of the process of solving MIPs. From a modeler's perspective, issues associated with the integer variables play a far greater role in determining the time required to solve the problem (or the quality of the solution found) than issues related to solving the LP. Hence, we focus our attention on understanding the process of solving a MIP. An understanding of this process can be very helpful in creating models that can be solved in a reasonable amount of time on a computer.

An important concept for solving MIPs is the notion of an LP *relaxation* of the MIP. One *relaxes* the requirement that the integer variables be integers and simply replaces it with a requirement that they be between their bounds (e.g., between zero and one). The LP relaxation for **mrp** is given as **relaxed mrp**.

$$\text{minimize: } \sum_{i=1}^{P}\sum_{t=1}^{T}(T-t)x_{i,t} \qquad \textbf{(relaxed mrp)}$$

subject to:

$$\sum_{\tau=1}^{t-LT(i)} x_{i,\tau} + I(i,0) - \sum_{\tau=1}^{t}\left(D(i,\tau) + \sum_{j=1}^{P} R(i,j)x_{j,\tau}\right) \geq 0 \qquad i=1,\ldots,P,\ t=1,\ldots,T$$

$$x_{i,t} - \delta_{i,t}LS(i) \geq 0 \qquad i=1,\ldots,P,\ t=1,\ldots,T$$

$$\delta_{i,t} - \frac{x_{i,t}}{M} \geq 0 \qquad i=1,\ldots,P,\ t=1,\ldots,T$$

$$\delta_{i,t} \geq 0 \qquad i=1,\ldots,P,\ t=1,\ldots,T$$

$$\delta_{i,t} \leq 1 \qquad i=1,\ldots,P,\ t=1,\ldots,T$$

$$x_{i,t} \geq 0 \qquad i=1,\ldots,P,\ t=1,\ldots,T$$

We have replaced $\delta_{i,t} \in \{0,1\}\ \ i=1,\ldots,P,\ \ t=1,\ldots,T$ by $\delta_{i,t} \geq 0\ \ i=1,\ldots,P,\ \ t=1,\ldots,T$ and $\delta_{i,t} \leq 1\ \ i=1,\ldots,P,\ \ t=1,\ldots,T$. This LP is not, in and of itself, a valid model. That is, the lot sizing constraints do not make sense unless the δ variables are either zero or one. However, if one were lucky the optimal solution to the problem called "**relaxed mrp**" would happen to have the property that all of the δ variables were either zero or one. For other models, not so much luck is required so it can happen that the solution to the LP relaxation is also the solution to the MIP.

Consider MIPs with minimization objectives. For every such problem, the objective function value for the optimal solution to the LP relaxation has to be as low or lower than the best possible solution to the MIP. This is because the best possible solution to the MIP is also a possible solution for the relaxation. Obviously, the reverse is not necessarily true because the best solution to the LP relaxation might give fractional values to some variables that must be integer valued in the MIP. We refer to the objective function value for the relaxation as a *lower bound* on the objective function value for the corresponding MIP. In the next section we describe how lower bounds are used in an algorithm for solving MIPs.

8.2 Branch and Bound

Branch and bound algorithms for binary MIPs work by fixing the values of some (or all) of the binary variables based on an educated guess concerning good values and then solving the corresponding relaxation. The algorithm begins by solving the LP relaxation for the original MIP. If there is no feasible solution, then there is no feasible solution for the MIP so the algorithm terminates. If the solution to the LP relaxation happens to have the property that all of the binary variables have values of either zero or one, then this is the optimal solution to the MIP, so the algorithm terminates.

Record keeping concerning the value of binary variables that have been tried is done using a tree data structure. The root of the tree is the LP relaxation, where none of the binary variables have been fixed at a value. We call the LP relaxation the *root node* of the tree. After the relaxation has been solved a variable is selected for *branching* along with a *branching direction*, which is said to be "up" if the variable is fixed at one and "down" if it is set to zero. This will constitute the first branch in the *branch and bound tree.* The branch will lead to a new node in the tree.

The root node is an LP relaxation of the original MIP. The first node in the tree is also a relaxation of a MIP, which is the original MIP except that one of the binary variables has been fixed at either zero or one. Suppose that we were solving the problem **mrp** and the first branch happened to be to set the value of $\delta_{2,3}$ to be one. The resulting relaxation would be

$$\text{minimize:} \quad \sum_{i=1}^{P}\sum_{t=1}^{T}(T-t)x_{i,t}$$

subject to:

$$\sum_{\tau=1}^{t-LT(i)} x_{i,\tau} + I(i,0) - \sum_{\tau=1}^{t}\left(D(i,\tau) + \sum_{j=1}^{P} R(i,j)x_{j,\tau}\right) \geq 0 \qquad i=1,\ldots,P,\ t=1,\ldots,T$$

$$x_{i,t} - \delta_{i,t}LS(i) \geq 0 \qquad i=1,\ldots,P,\ t=1,\ldots,T$$

$$\begin{array}{lll} \delta_{i,t} - \frac{x_{i,t}}{M} & \geq 0 & i = 1, \ldots, P, \ t = 1, \ldots, T \\ \delta_{i,t} & \geq 0 & i = 1, \ldots, P, \ t = 1, \ldots, T \\ \delta_{i,t} & \leq 1 & i = 1, \ldots, P, \ t = 1, \ldots, T \\ x_{i,t} & \geq 0 & i = 1, \ldots, P, \ t = 1, \ldots, T \\ \delta_{2,3} & = 1 & \end{array}$$

This problem asks the question "what is the best that can happen if we require production of SKU 2 in period 3?" If the LP solver reports that there is no feasible solution, then there can be no feasible solution with $\delta_{2,3} = 1$ for the MIP, so the node is deleted from the tree. If there is a feasible solution, the node is kept in the tree. For each node the tree stores information about which variables are fixed at which values, the objective function value for the corresponding LP relaxation, and perhaps some other information to improve efficiency.

Any of the binary variables whose values have not been fixed by the branching process at a node are said to be *free* at that node. In our example, all of the δ variables except for $\delta_{2,3}$ are free at the first node. Each branch from a node is labeled with the name of the branching variable and the direction. So for our example, the first branch would be labeled ($\delta_{2,3}$, up). When a node is deleted from the tree (for example, if its relaxation is infeasible), the branch information is kept. At each node, the variables for which there is only one branch will be referred to as *partially explored* because both of their possible values have not been tried yet. For our example, the variable $\delta_{2,3}$ would be partially explored after the first iteration.

At each *iteration* of the branch and bound process a node is selected. Then a variable is selected from among the free and partially explored variables at the node. If it is a free variable a branching direction is selected. If it is a partially explored variable the branching direction is clearly the direction that has not yet been tried. The information stored for the selected node, along with the selected variable and branching direction give rise to a new MIP and a new node in the branch and bound tree with one more variable fixed than for the selected node. The resulting relaxation is solved or deemed to have no feasible solution (LP software can do both at the same time). If it has no feasible solution it is deleted.

If the MIP has any feasible solutions, eventually this algorithm will find one. The methods of selecting nodes and variables is beyond the scope of this book, but the reader can agree that any method will eventually lead to trying all combinations of binary variables that cannot be ruled out by the fact that they are part of an infeasible relaxation. In the worst case, every combination of binary variables will be tried.

For most feasible problem instances, the branch and bound algorithm will encounter a solution that is feasible for the MIP. It could be that a node in the tree is reached where all of the integer variables are fixed and the resulting relaxation is feasible, or it could be a node where some of the integers are fixed and the others take on integer values in the optimal solution to the relaxation. Either way, the first integer feasible solution that is encountered provides an *upper bound* on the best feasible objective value. It is an upper bound because obviously the optimal solution can be no worse than this solution. If another solution is found that is feasible for the MIP it will become the new upper bound if its objective function value is better than the current upper bound.

Once an upper bound is known, all those nodes in the tree with a relaxation objective function value that is worse than the upper bound can be deleted. There is no way that they can result in a node or a relaxation that will be better than the current upper bound, hence they cannot result in discovery of an optimal solution. All of the nodes that are created from such a node will have the same fixed variables as the node and more, but fixing more variables cannot improve the objective function value since they will not be fixed at values outside the values that they are allowed to take in the relaxation.

The ability to delete nodes can have a dramatic impact on the solution time. Consider a problem with 1000 binary variables. If absolutely none of the nodes can be deleted during execution of the algorithm, then about 2^{1000} relaxations must be solved. Even if relaxations can be solved quickly, they do require non-trivial effort. Node deletion can reduce the number of nodes dramatically. When a node is deleted, all of the nodes that would have emanated from it will not even be created as a result. If the node is near the root node and deleted early in the process the number of nodes that are ultimately created is reduced significantly.

If the branch and bound algorithm has no nodes left with free variables, it cannot be expanded further and the solution corresponding to the upper bound is clearly an optimal solution. If there is no such solution, then the problem clearly has no feasible solution. In both cases, we say that the problem has been completely *solved* or sometimes we say that it has been solved to provable optimality (or that infeasibility of the problem instance has been proven). Some practical problems with nearly 1000 integer variables can be completely solved with only about 100,000 nodes or fewer. But for some others, the tree keeps growing beyond a million nodes with no feasible solution found. In such unfortunate situations, we cannot rule out the existence of a feasible solution, but we cannot find one in a reasonable amount of time either.

A large number of MIPs are of an intermediate variety. For these problems, some feasible solutions can be found for the MIP, but not enough of them and not early enough to prevent the tree from growing and growing until it has exceeded reasonable computer memory. In addition to exceeding memory limits, such problems can execute for days on a large computer and the tree can still be growing. However, if the branch and bound algorithm has found some feasible solutions, these can be used even though we cannot be sure that there is not a better solution possible.

Having a solution that is proven to be optimal because a branch and bound tree has exhausted all possibilities either through branching or node deletion is good, but usually not essential. These models are, after all, just approximations to the goal of producing good production plans. The solution that corresponds to the minimum objective function value might not really be that much better than some other feasible solutions that have objective function values that are almost as low.

There are many ways to make the branch and bound algorithm more efficient, but all other things being equal, having more integer variables means that more nodes will be required to solve the problem and more nodes will be required to find a feasible solution. Commercial solvers are continually being improved, but some care must be taken to avoid models that are too big to solve. In §6.6 we describe methods of reducing the number of resources in the model and methods of combining SKUs so as to reduce the number of decision variables. In §8.4 methods that look for good, feasible solutions but that cannot prove optimality are described. These methods can often find feasible solutions more quickly than branch and bound, particularly when there are some non-linear elements in the model. Before proceeding with these methods that change the model or the solution method, we describe some special variables that can reduce the size of the relaxations and/or the branch and bound tree.

8.3 Special Variable Types

There is a number of special variable types, but we will consider just those that can have the greatest impact on the models that we are developing for supply chain planning. Not all types of special variables are supported by all commercial solver software. None of these variables are truly necessary in the sense that the same thing can be accomplished with combinations of variables that are either real valued or binary. However, they can make the models easier to solve and in some cases simpler and thus easier to read.

8.3.1 Semi-continuous Variables

The first type of variables that we will consider are referred to as *semi-continuous* variables. A semi-continuous variable must be either zero or any value above a specified limit other than zero. Such variables are ideal for modeling production quantities that are subject to a minimum lot size constraint. In the original problem **mrp**, the δ variables could be eliminated from the model if the x variables for products with a minimum lot size were semi-continuous.

Branch and bound algorithms can branch on semi-continuous variables in a fashion that is very similar to the branching process for binary variables.

When the branching direction is "down," the variable is forced to be zero. When the direction is "up," the variable is constrained to be above the specified limit. The branch and bound tree will generally not be any smaller, but the relaxations will be a little bit easier to solve because constraints that define the relationship between the binary variable and a real valued variable will not be needed. This relationship is subsumed by the semi-continuous variable definition.

We adopt the convention that a variable is declared to be semi-continuous by using a double colon followed by the non-zero limit. So, for example, the constraint $x_{1,2} :: LS(1)$ indicates that the variable $x_{1,2}$ must be either zero or any value not less than $LS(1)$. We adopt the convention that if the non-zero limit is zero, then the variable can take on any value zero or greater and there is no need to branch on it. The problem **mrp** can be rewritten using semi-continuous variables as follows:

$$\text{minimize: } \sum_{i=1}^{P}\sum_{t=1}^{T}(T-t)x_{i,t} \qquad \textbf{(semi-cont mrp)}$$

subject to:

$$\sum_{\tau=1}^{t-LT(i)} x_{i,\tau} + I(i,0) - \sum_{\tau=1}^{t}\left(D(i,\tau) + \sum_{j=1}^{P} R(i,j)x_{j,\tau}\right) \geq 0 \qquad i = 1,\ldots,P,\ t = 1,\ldots,T$$

$$x_{i,t} \quad :: LS(i) \qquad i = 1,\ldots,P,\ t = 1,\ldots,T$$

This model is much easier to understand than the original mrp model. It is also somewhat easier to solve because branching is only necessary once for each variable in the set.

8.3.2 General Integer Variables

We have argued that in most planning environments, production quantities can reasonably be allowed to take on real values even if that is not reasonable for scheduling. The idea is that the fractional portions in the production plan constitute a small error that is usually dwarfed by estimation errors associated with the data. Capacity data are particularly difficult to estimate with extreme precision in most production environments.

However, there are circumstances where integrality is too important to ignore. Many of the simpler formulations might result in integers with no special effort, but for some formulations there will be production quantities that take on fractional values that are not reasonable. One can declare such variables to be *general integer* variables. These variables are required to take an integer value between a lower and an upper limit. The branch and bound algorithm for general integers is similar to the binary version and the branching directions generalize easily. Some additional record keeping is required and the number of possible branches goes up rapidly with the difference between the lower and upper limit.

General integers could also be used to obtain fixed lot sizes, rather than minimum lot sizes. With a general integer variable, $\gamma_{i,t}$ constrained to be an integer greater than or equal to zero, we can then constrain production for an SKU i using

$$x_{i,t} = \gamma_{i,t} LS(i)$$

for some or all time buckets t. Of course, if the constraint is for all time buckets, then there is no need for general integers. The production quantities can simply be rescaled to be in multiples of the lot size.

8.3.3 Special Ordered Sets

Special ordered sets were developed to allow modelers to communicate important information to the branch and bound algorithm. The effects provided by these special types of variables can be obtained using simple binary variables, but the resulting models are harder to read and require more computer time to solve. There are two types of special ordered sets, creatively named type one and type two. They are referred to as SOS1 and SOS2, respectively. Special ordered sets are a list of variables. In the case of SOS1, exactly one of the values can be non-zero. In the case of SOS2, at most two values in the list can be non-zero and if there are two, they must be adjacent members of the list.

SOS1. Variables declared to be SOS1 are used for multiple choice selection. An example of a multiple choice situation is when there is a group of products that have the property that if one of them is produced in a time bucket the others cannot be produced. The δ variables for these SKUs could be collected in a special ordered set of type 1. A separate set would be needed for each time bucket.

SOS2. Variables declared to be SOS2 are used to create piecewise linear approximations to non-linear functions. In actuality, of course, almost nothing is truly linear but linear approximations are often reasonable. When linear approximations are not good enough but piecewise linear approximations are, then SOS2 variables can be used. In §4.4.3, we mentioned the possibility of using non-linear approximation to the utilization by parts that require a significant changeover on a resource composed of parallel machines. For the resources and SKUs to be modeled in this way, we fit a curve and then produce a piecewise linear approximation as shown in Figure 8.1. Rather than using a linear function such as $U(i,k)x_{i,t}$ for the fraction of resource k utilized by SKU i in time bucket t, we can use the piecewise linear approximation given by the figure. For any production quantity measured along the horizontal coordinate axis, we read the corresponding utilization fraction on the vertical coordinate axis. The utilization per piece goes down as the production quantity goes up due to amortization of setup and changeover times across a larger production volume. Such a shape could also be due, in part, to learning curve effects. Since the graph in Figure 8.1 gives the utilization as a function of production quantity, the slope of the line segments gives the utilization per piece. As the quantity increases, each new line segment has a lower slope.

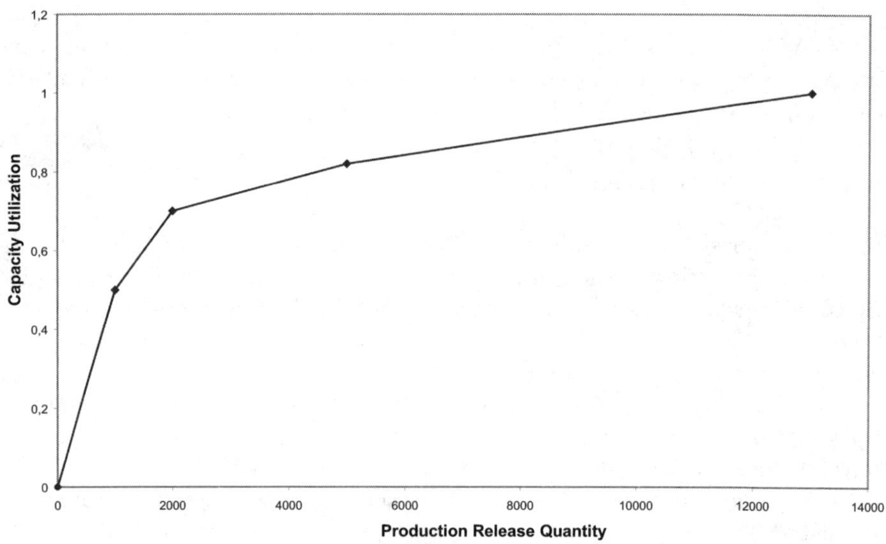

Fig. 8.1. Piecewise Linear Approximation to the Utilization Fraction as a Function of the Production Quantity

For this example, we are assuming that there is only one such SKU and one such resource, so the breakpoints have only one index. If there were multiple SKUs and/or multiple effected resources, then there would need to be breakpoints for each and the respective indexes would have to be added to the breakpoint data. For the discussion of breakpoints, we temporarily introduce the notation X for the list of production quantity breakpoints and Y for the corresponding utilization breakpoints. We refer to the number of breakpoints as B. The actual values for $X(1), X(2), \ldots, X(B)$ and $Y(1), Y(2), \ldots, Y(B)$ would have to be determined by measuring a curve giving the utilization of the resource as a function of the production quantity which would have to be obtained using historical data or engineering estimates.

If we let the variable $u_{i,t}$ take the place of $U(i,k)x_{i,t}$ for SKU i on resource k (suppose i is 1, then the capacity constraint for resource k would be

$$u_{1,t} + \sum_{i=2}^{P} U(i,k)x_{i,t} \leq 1$$

or something similar) we can compute its value using variables of type SOS2. Let the variable set λ_ℓ, $\ell = 1, \ldots, B$ be SOS2, then we add the constraints

$$
\begin{array}{lll}
x_{i,t} & = \sum_{\ell=1}^{B} X(\ell)\lambda_\ell & t = 1, \ldots, T \\
u_{i,t} & = \sum_{\ell=1}^{B} Y(\ell)\lambda_\ell & t = 1, \ldots, T \\
\lambda_\ell & \geq 0 & \ell = 1, \ldots, B \\
\lambda_\ell & \leq 1 & \ell = 1, \ldots, B \\
\sum_{\ell=1}^{B} \lambda_\ell = 1 & & \\
\lambda & & \text{are SOS2}
\end{array}
$$

to give x and u the appropriate values. The fact that at most two λ variables may be non-zero and if there are two, they must be adjacent will cause the u and x values to correspond according to a line segment in the piecewise linear function defined by the X and Y values. For example, if Figure 8.1 were used to provide breakpoints, they would be where the slope changes. Modeling languages such as Mosel have a more compressed syntax for SOS2 constructs, but the concepts are the same.

Exact methods of solving mixed-integer programs can be effective for many kinds of programs. If there are a large number of non-linearities that are important to the model, or too many integers, then heuristic search methods must be considered. This is the subject of the next section.

8.4 Heuristic Search Methods

We must pause briefly for a technical issue. It is useful to make a distinction between *exact* optimization methods that are guaranteed to find an optimal solution if one exists and *heuristic* methods that tend to look for good solutions, but generally offer no practical guarantees concerning performance. Branch and bound is an exact method. Obviously, all other things being equal, one would prefer an exact method. However, for many large problems the requirements for time and/or computer memory force the use of a heuristic or termination of exact methods before they have found a solution that is optimal.

Many problems associated with production planning and scheduling in a supply chain are too large, have too many integers and/or have portions that are highly non-linear so that commercially available MIP solvers cannot be used to solve them. Under these circumstances, branch and bound algorithms must be augmented by or replaced by heuristic methods. One important class of heuristic methods is based on meta-heuristic search strategies.

Meta-heuristics provide a framework for optimization that relies on some components that are specific for the problem being solved.

These methods offer the advantage that they can address a broad range of problem types, including those with non-linearities and a large number of integer variables. This flexibility comes at a cost. The first cost is that there are not readily available interfaces to modeling languages. Commercial, well-supported MIP solvers have interfaces to commercially available well-supported modeling languages. These modeling languages allow entry of the models that we have developed in a form that is very similar to what you see in this book. The range of problems addressed by heuristic search methods is simply too large to facilitate use of an algebraic language. Typically, the "interface" to heuristic search methods is a programming language. That is, the problem definition must be done by writing subroutines to read the data and evaluate the objective function, etc. This is changing. At the time of this writing, some special purpose interfaces exist as well as some graphical interfaces. Another drawback is that at the time of this writing, there are very few commercial codes to support heuristic search. The major ERP vendors and supply chain management software vendors produce their own heuristic search code and others do the same or else make use of portions of academic codes. The final drawback is that the modeler typically must provide some assistance in the solution methods that is provided in the form of subroutines to evaluate *neighborhoods*. This will be clarified shortly.

An important use of heuristic search methods in planning is to verify the existence of a feasible schedule that corresponds to a potential plan. As we have noted, scheduling problems are often fairly complicated and involve too many integer variables and non-linearities to be solved using a MIP solver. This is particularly the case when there are significant changeover times that are a complicated function of the production sequence. A reasonable procedure is to make simplifying assumptions and perhaps perform aggregation to obtain a solvable planning model. Once a plan has been created (and perhaps disaggregated) it implies scheduling problems for critical resources.

That is to say, the question arises as to whether a feasible schedule can be found that will result in producing the quantities given by the plan. Heuristic search methods can be employed to seek a solution. If none can be found, then the utilization data or the changeover time data (or both) can be changed (i.e., increased) and a new plan can be found with lower production quantities for which a feasible schedule might be obtained. Heuristic search methods are extremely general and, therefore, can be applied to almost any conceivable scheduling problem.

8.4.1 A Brief Primer on Heuristics

Heuristic methods provide simple means of indicating which among several alternatives seems to be the best; and basically they are based on intuition. Be sure to reread the words "seems to be" and remember that we distinguished between heuristic methods and exact optimization. That is, heuristic methods may be seen as criteria, methods, or principles for deciding which among several alternative courses of action promises to be the most effective in order to achieve some goal. In a sense they may be regarded as rules of thumb representing compromises between the requirements to make such criteria simple and the desire to see them discriminate correctly between good and bad choices.

One of the buzz words often related to heuristic methods is that of greediness. *Greedy heuristics* usually work in an iterative fashion with some kind of myopic behavior. Assume that we want to "construct" some solution (for instance, for **mrp**). In each iteration there is a number of alternative choices that can be made. From these alternatives which consist in fixing one or more variables, a greedy choice is made, i.e., the best alternative according to a given evaluation measure or simply according to some rule of thumb is chosen until we have obtained a feasible solution. That is, usually, a greedy construction heuristic starts with an incomplete solution and completes it iteratively.

8.4.2 Abstract Formulation and Solution Representation

In order to further understand heuristic search methods as general purpose methods, we need to consider an extremely abstract problem formulation. We need a formulation that is so abstract that it is general enough to capture the range of problems that can be addressed using heuristic search. A brutally abstract formulation, and the one that we will use, is simply to

$$\text{minimize} \quad f(\nu)$$

were ν is called the decision vector. We use the symbol ν to summarize all of the variables over which we might need to search. The function f captures the objective function and the constraints. It returns the objective function value if ν is a feasible solution and a very large value if ν is infeasible. We construct the function f so that it takes on extremely large values if ν does not satisfy problem constraints.

As an example, consider the problem **mrp**. The vector ν could be a list of the x and δ values. To compute f, one checks to see if the constraints are met by these values and if so, the function is the same as the objective value. If not, the function takes a large value (we have called such a value, M) or it might be M plus the number of constraints that are not satisfied. This scheme makes it so that we can distinguish between infeasible solutions, but no infeasible solution could be considered better than any feasible solution since M is chosen so as to be "very large." In this case that means we must pick a value for M that is larger than any possible feasible objective function value.

Another alternative for problems such as **mrp** is to use just the δ values for the vector ν. This makes the ν vector much shorter at the expense of making the f function more complex. Under this scheme, to compute the f function for a particular ν vector, one would need to fix the δ values and then use an LP solver for the resulting problem. If the problem was feasible, f would be the objective function value for the optimal solution to the LP. If the LP solver reported that no feasible solution exists, then f would take the value M. This scheme can be summarized as

$$f(\nu) = \text{minimize:} \quad \sum_{i=1}^{P}\sum_{t=1}^{T}(T-t)x_{i,t} \qquad \textbf{(mrp)}$$

subject to:

$$\sum_{\tau=1}^{t-LT(i)} x_{i,\tau} + I(i,0) - \sum_{\tau=1}^{t}\left(D(i,\tau) + \sum_{j=1}^{P} R(i,j)x_{j,\tau}\right) \geq 0 \qquad i=1,\ldots,P,\ t=1,\ldots,T$$

$$x_{i,t} - \nu_{i,t}LS(i) \geq 0 \qquad i=1,\ldots,P,\ t=1,\ldots,T$$

$$\nu_{i,t} - \frac{x_{i,t}}{M} \geq 0 \qquad i=1,\ldots,P,\ t=1,\ldots,T$$

$$x_{i,t} \geq 0 \qquad i=1,\ldots,P,\ t=1,\ldots,T$$

$$\nu_{i,t} \in \{0,1\} \qquad i=1,\ldots,P,\ t=1,\ldots,T$$

which would only be useful if ν had binary values. The constraint $\nu_{i,t} - \frac{x_{i,t}}{M} \geq 0$ would force production of SKU i to be zero in period t if $\nu_{i,t}$ were zero and would allow production if it were one (but production would not be required, just allowed). This *solution representation* scheme offers advantages for the methods that we will describe. As a point of philosophy, we note that it is also the representation used by branch and bound.

For sequencing problems, such as those that often arise in detailed scheduling, it is usually better to use a direct representation of the sequence rather than a large, sparse matrix of binary variables. To discuss sequencing it is convenient to speak in terms of *jobs.* Jobs are a collection of one or more of the same SKU to be produced. In the simplest case, each job consists of one piece. However, it often makes sense to collect into a job the pieces that correspond to a customer order or that will be transported together.

A sequence can be represented directly in a variety of ways. One way is to use a vector – call it σ – for which each index corresponds to a position in the sequence and the value of the vector element indicates the job that occupies that position in sequence (sometimes also called permutation). Using this representation, if $\sigma_j = i$ then the jth part to be produced is SKU i. If the vector is printed, the sequence of jobs can be read directly. We will refer to this as the *direct sequence* representation. An alternative representation uses a vector – call it α – so that the indexes correspond to job numbers and the values correspond to positions in the sequence. So $\alpha_i = j$ means the same thing as $\sigma_j = i$. We refer to this as the *positional sequence* representation.

As an example let us assume the production of four jobs. Then producing the jobs in the sequence job 1 first, then job 4, then job 2, and finally job 3 results in $\sigma = (1, 4, 2, 3)$ and $\alpha = (1, 3, 4, 2)$, respectively.

8.4.3 Example of an Embedded Problem

To create a concrete example, consider a situation where there are no out-of-pocket costs for a changeover and no wasted material so C and W are zero. However, suppose further that a significant amount of time is needed to change over some bottleneck resource, k', and this time depends on both the current job and the next job. We will reuse the symbol S for these changeover data. For simplicity, assume that each job is one piece. In such a situation, we might make use of a direct sequence representation as a basis for heuristic search.

In order to write the model in a concise way, we will make use of the fact that we can go back and forth between the familiar quantity representation, x_{it}, and a sequence representation where σ_j gives the job number of the jth job. We could modify the model **SCPc** as follows: For resource k' for each time t replace the constraints

$$
\begin{aligned}
\sum_{i=1}^{P} [U(i,k)x_{i,t} + S(i,k)\delta_{i,t}] &\leq 1 + y_{k,t} \\
y_{k,t} &\leq F(k,t) \\
y_{k,t} &\geq 0
\end{aligned}
$$

with the embedded problem

$$\text{minimize: } y_{k',t}$$

subject to:

$$
\begin{aligned}
\sum_{j=1}^{n} U(\sigma_j, k') + S(\sigma_{j-1}, \sigma_j, k') &\leq 1 + y_{k',t} \\
y_{k',t} &\leq F(k',t) \\
y_{k',t} &\geq 0
\end{aligned}
$$

The data $S(\sigma_{j-1}, \sigma_j, k')$ could be an array giving the changeover times or a complicated function. For this to make sense, the value of σ_0 must be given as data or an assumed value must be used.

The main thing to notice is that constraints have been replaced with an optimization problem. In doing so, we have embedded a higher level of scheduling detail in the planning model. This modification will result in a model that will check for the existence of a feasible sequence that corresponds to the plan while at the same time seeking a good sequence. For a general changeover matrix or function, the sequencing problems are too hard to solve to proven optimality.

For such embedded problems, heuristic search methods are appropriate. These methods offer a great deal of flexibility, e.g., the changeover time can be a function rather than data at the expense of not finding a solution that is provably optimal. The flexibility also makes it possible to solve models that have non-zero C and W cost data, but the statement of the model is a little more complicated. In order to proceed with our discussion of heuristic search methods, we describe their main components.

8.4.4 Neighborhoods and Evaluation Functions

The methods that we shall describe rely on a *neighborhood* structure and an *evaluation* function. Both can be, or must be, designed as part of the process of implementation on a computer. Furthermore, both of these things can make use of cleverness supplied by the modeler, which is why we describe them here.

A neighborhood is based on the idea of *moves* that begin with one solution representation, ν, and transform it into a different solution, ν'. Algorithms that move iteratively from one solution to the next are called *local search* algorithms. We refer to solution ν' that can be reached from ν using a single move as a *neighbor* of ν. We refer to all such solutions as being in the *neighborhood* of ν. For the purpose of discussing local search, let $\mathcal{N}(\nu)$ denote the set of solutions that can be reached from ν with a single *move.* In other words, $\mathcal{N}(\nu)$ is the neighborhood of ν. We refer to the solution ν that defines it as the *center* of the neighborhood. To indicate the number of members in a neighborhood, we use the notation $\mid \mathcal{N}(\nu) \mid$.

Unless stated otherwise, we will be referring to *perturbation* or *transformation* moves, that operate on full solution vectors. In contrast, *construction* moves operate on a partial vector (one where not all values in ν have been specified) and result in a vector of increased length.

For problem **mrp**, if ν is equivalent to the δ variables, we could imagine using any one of a variety of neighborhood structures. The simplest would be a neighborhood based on moves that involve changing the value of one element of ν. So to create a neighbor of a solution, we would take one value of $\delta_{i,t}$ and change it (if it was zero, it would become one and if it was one, it would become zero). The size of the corresponding neighborhood is P times T, since each neighbor is defined by one element of δ changing values. To have a name for this neighborhood structure, we will call it a "bit-flip" neighborhood.

To describe some of the range of possibilities for neighborhood design, consider the sequence representations given above. We can perform a *swap* move, where one vector element is exchanged with the other. In the case of the direct sequence representation, such a move can be visualized as two people (the jobs) trading positions in a waiting line (the sequence). In the case of the positional representation, such a move corresponds to two positions trading jobs. An *insertion* move is easy to visualize in the case of the direct sequence representation: A person leaves her place in line and "butts in" somewhere else (perhaps further back in the line).

We may think of the sequences as forming a graph with the nodes of the graph representing the jobs and the edges of the graph representing the order of the jobs. With this solution representation, we can make use of *k-opt* moves. Figure 8.2 (a) gives a sequence with nine jobs as an example. Figure 8.2 (b) and (c) show the outcome of a 2-opt move and a 3-opt move. Such moves are the result of exchanging k edges in the graph (where k is equal to two and to three in our example). While k edges have been removed from the original sequence, some other k edges (shown as dotted lines) have been included respectively.

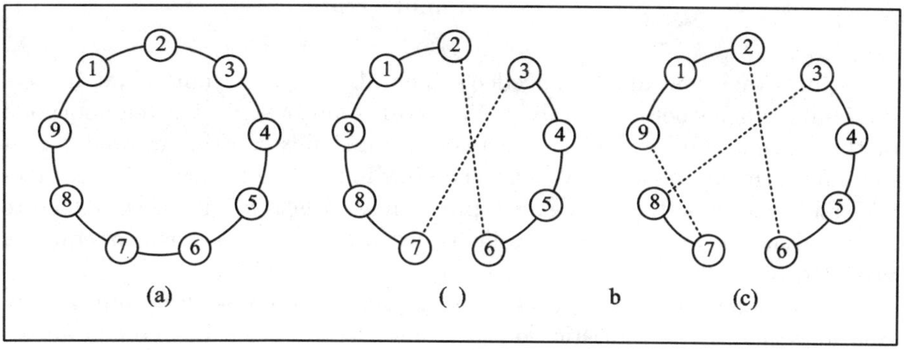

Fig. 8.2. k-opt Moves

To complete our discussion of neighborhoods, we consider one that is of both theoretical and practical interest. An *enumeration* neighborhood has the property that every feasible solution can be reached from every other feasible solution in one move. So for a sequencing problem using the direct sequence representation, the neighborhood of a particular solution consists of every permutation of the jobs minus the one particular permutation that is the center of the neighborhood. Of course, for a sequencing problem, one would never try to explore every solution. Such neighborhoods are used only in conjunction with algorithms that select neighbors at random.

Another property of theoretical and practical interest in neighborhood design is whether the neighborhood is *connected*. This refers to the useful property that every feasible solution can be reached from every other feasible solution in a sequence of moves (not necessarily only one). Clearly, enumeration neighborhoods are an example of connected neighborhoods.

A function, $\hat{f}(\nu)$, must be specified for move evaluation. The function usually resembles f, but may differ so it is easier to compute or in order to take advantage of special properties of the problem and/or the neighborhood structure. For our purpose, one can think of the evaluation function as being an approximation to f. In many applications $\hat{f}$ is f (in which case, the approximation is quite good).

A simple approach to addressing optimization problems is referred to as *steepest descent.* Such an algorithm begins with an initial solution as the *current solution* and then picks the neighbor with the best evaluation function value to become the new current solution. This algorithm proceeds iteratively until there are no improving moves in the neighborhood of the current solution. If all solutions in the neighborhood are worse than the current, such a solution is referred to as a *local minimum* with respect to the neighborhood and evaluation function.

We can restate this algorithm using abstract notation as follows. A steepest descent algorithm begins with an initial solution, $\nu^{(0)}$, and selects solution $\nu^{(i)}$ at iteration $i > 0$ using the relation

$$\nu^{(i)} \leftarrow \operatorname*{argmin}_{\nu \in \mathcal{N}(\nu^{(i-1)})} \hat{f}(\nu) \tag{8.1}$$

(a tie breaking rule may be needed). The algorithm terminates at a local minimum when there are only higher cost solutions in the neighborhood of the current solution. The notation "argmin" means "the argument that minimizes the expression." We use the notation $x \leftarrow y$ to mean that x takes the value held by y. Read such an expression as "x gets y." This is intended to show a dynamic assignment of value as distinct from a statement of equality provided by $x = y$.

We consider steepest descent to be a heuristic method of finding a solution. It relies on the heuristic notion of improvement, but has no guarantees of finding the optimal solution. The methods we will consider are broadly grouped under the title *heuristic search* and the title *meta-heuristic.* The prefix "meta" refers to the fact that these methods provide general-purpose control mechanisms for a search based on an underlying heuristic such as local search.

The first meta-heuristic that we will consider is referred to as *iterated local search* or ILS. ILS performs a steepest descent using some neighborhood structure and when a local minimum is reached, some number of so-called *kick moves* are performed using a different neighborhood structure (or occasionally just a different evaluation function). In this way, the search escapes the local minimum and can begin a descent using the original neighborhood structure. For example the algorithm might use 2-opt moves until a local minimum is reached and then perform a few k-opt kick moves with $k > 2$, where the meaning of "a few" and the values of k are controlled by parameters and may vary during the search. After the kick moves, the search resumes with 2-opt moves beginning with the solution that results from the kick moves. There are many variants of ILS, but this introduction gives the reader an overview of the method.

A simple way of escaping local minima is to restart the steepest descent at a random or arbitrary solution after a local minimum has been reached. We refer to this algorithm as steepest descent with random restarts (SDRR). The meta-heuristic SDRR can be viewed as ILS with a single kick move that uses an enumeration neighborhood, a random neighbor selection mechanism and an evaluation function that accepts any move it sees. There are many other ways of using neighborhoods to look for good solutions for hard problems, but SDRR is actually often better than most of them. However, other methods are very popular for a variety of reasons so we proceed with a brief description of some of them.

Using randomness in a slightly different fashion is done in an algorithm called greedy randomized adaptive search procedure (GRASP). Assume a greedy heuristic. If we omit a greedy choice criterion for a random strategy we can run the algorithm several times and obtain a large number of different solutions. Thus a combination of best and random choice seems to be appropriate. Consider a list consisting of a number of (best, i.e., first best, second best, third best etc.) alternatives and call it a candidate list. The length of the candidate list is given either as an absolute value or a percentage of all feasible alternatives. At the very heart of GRASP a number of alternatives out of this candidate list is chosen randomly and a local search procedure is applied to each of them and the best found solution is taken. The underlying principle of GRASP is to investigate many hopefully good starting points through the local search procedure and thereby to increase the possibility of finding a good local optimum on at least one replication. The method is said to be adaptive as the greedy function takes into account previous decisions when performing the next choice.

It is necessary for modelers to understand the basics of heuristic search methods because in order for these methods to be successful, modelers must be involved in the process of defining neighborhoods and evaluation functions. In subsequent sections we will cover the fundamental concepts necessary for the creator of a model to provide assistance to the people who are doing algorithm development.

There are a large number of meta-heuristics in the academic literature, but the most popular methods based on local search are simulated annealing and tabu search. So-called genetic algorithms were forced to undergo a name change when it was discovered that they did not perform well unless they contained a significant local search component. These algorithms are now called evolutionary algorithms, and they make use of populations of solutions, but typically also perform local search. All of these methods provide mechanisms for escaping local minima.

In spite of various mathematical theorems concerning stylized versions of these algorithms, there are no performance guarantees in practice. Often, a large production planning problem, or even a fairly small production sequencing problem cannot reasonably be solved to provable optimality anyway. These methods offer the advantage that they can explore solutions one after the other and might (and often do) encounter good solutions along the way.

In each of our algorithm descriptions, we include a function named BestCheck that compares the current solution with the best found so far and updates the best found as appropriate. General purpose functions are summarized in Table 8.1.

Name	Description
BESTCHECK(ν)	If ν is better than the best seen so far, update the best seen so far.
URAN(l, u)	Return a random number between l and u.
IRAN(l, u)	Return a random integer between l and u.
Obj(ν)	Value of the evaluation function for ν.

Table 8.1. General Purpose Functions Used in Algorithm Statements

8.4.5 Simulated Annealing

Simulated annealing (SA) is based on an analogy with cooling metal. The idea is that when metal cools, the molecules perform in a way that can be modeled as a local search for the minimum energy configuration. The actual behavior of an SA algorithm has only a tenuous connection with actual behavior of cooling metal, but the metaphor is appealing. The other thing that is appealing to some people about SA is the existence of some mathematical theory concerning the behavior of some SA algorithms. The algorithms are not the same as SA in practice and the behavior is present only after an infinite amount of time. Furthermore, the theoretical behavior of the SA algorithms is dominated by the behavior of completely random search. However, the theory requires very sophisticated mathematics and is quite interesting in its own right. The best SA algorithms do often perform better than SDRR and have been effective in getting good solutions to hard problems.

SA algorithms escape local minima by using randomness in move selection. At each iteration of a simulated annealing algorithm, a move is selected at random and the change in cost is computed for the move. Let $\Delta = \hat{f}(\nu') - \hat{f}(\nu)$. If a move from ν to ν' is considered, it is accepted (the move is made) with probability

$$\exp(-\Delta/T)$$

if Δ is positive (the move not improving) or with probability 1 if Δ is not positive. We refer to T as the *temperature*. For high values of T, the acceptance probability is near one for all moves; hence, the algorithm proceeds from solution to solution at random. For very low values of T, only moves that result in improvement (a decrease in the cost function) are likely to be accepted. SA algorithms begin with high values of T and slowly (over a large number of iterations) lower the temperature. The exact specification of this process is referred to as a *cooling schedule.*

We have adopted the algorithm statement and parameterization given in Table 8.2 and Figure 8.3 because it is simple and although not necessarily the best, the algorithm performance is quite stable. The function InitializeTemp(INITPROB) uses a number of iterations at increasing temperatures in order to return a temperature value that will result in approximately INITPROB of moves being accepted. The function Frozen(MINPERCENT) returns true if five temperatures in a row result in less than MINPERCENT percent acceptance of moves.

Name	Description
INITPROB	The initial temperature is set so that the probability of accepting moves is approximately INITPROB.
TEMPFACTOR	The temperature is reduced by multiplying the current temperature by TEMPFACTOR.
SIZEFACTOR	The number of iterations at each temperature (both during InitializeTemp and cooling) is SIZEFACTOR times the neighborhood size.
MINPERCENT	The algorithm terminates when five temperatures in a row result in less than MINPERCENT percent acceptance of moves.

Table 8.2. Canonical SA: Parameters

```
Begin with a randomly selected solution ν
T ← InitializeTemp(INITPROB)
REPEAT
   REPEAT SIZEFACTOR × | N(ν) |  times
      Randomly select ν' ∈ N(ν)                              (a neighbor)
      Δ ← f̂(ν') − f̂(ν)                                      (change in Obj)
      IF (Δ ≤ 0) OR (exp(−Δ/T) < URan(0,1))                  (good enough?)
         ν ← ν'                                              (move)
         BestCheck(ν)
   T ← TEMPFACTOR × T                                        (cool)
UNTIL Frozen(MINPERCENT)
```

Fig. 8.3. Canonical Simulated Annealing

SA algorithms terminate naturally based on the value of MINPERCENT, but this can sometimes require far too much time. In such cases the algorithm must be simply stopped and the best value found so far must be taken as the best available answer. Basically the goal in setting SA parameters is to reduce the running time to a reasonable level without too much degradation in performance.

8.4.6 Tabu Search

Unlike simulated annealing, canonical tabu search (TS) is deterministic; although many implementations have probabilistic aspects. TS algorithms select moves according to a steepest descent scheme, except they make use of a *tabu list* to force the search away from solutions selected for recent iterations. This tends to cause the search to escape local minima. Moves are rejected if they satisfy conditions given by the tabu list, which in its simplest form can be thought of as a first-in-first-out (FIFO) list based on certain attributes of the most recent κ moves. Solution vectors that are the result of moves having the attributes found on the tabu list are removed from consideration unless they meet an *aspiration criterion.* Aspiration criteria are used to avoid removing very good moves from consideration. Since attributes of moves are tabu, rather than recent solutions, many trial solutions may be forbidden that have not been recently visited. Aspiration criteria are intended to make sure that obviously good moves are not avoided because of the tabu mechanism.

A simple TS is shown in Figure 8.4. The only parameter is TABUMULT which is used to compute the tabu list length, $\kappa \leftarrow$ TABUMULT $\times\, n$. Once a bit position has changed value, the tabu list forbids reversing the change for κ moves. The function Tabu(ν, ν') returns true if attribute(s) of the move is/are in the tabu list and the aspiration criteria are not satisfied. The function UpdateTabuList(ν,ν'') adds the attribute(s) of the move to the tabu list and deletes the oldest member of the list.

```
Begin with a randomly selected solution ν
REPEAT
   F ← +∞
   FOR each ν' ∈ N(ν)                              (a neighbor)
      IF Obj(ν') < F                       (best move so far?)
         IF NOT Tabu(ν, ν')                              (OK ?)
            F ← Obj(ν'); ν'' ← ν'                 (record best)
   UpdateTabuList(ν, ν'')                     (forbid reversal)
   ν ← ν''                                     (make the move)
   LTMem(ν)
   BestCheck(ν)
UNTIL TimeLimit()
```

Fig. 8.4. Canonical Tabu Search

Setting the parameter values for a simple TS is particularly easy since there is only one parameter and performance is fairly robust to the choice of this parameter. The parameter is the tabu list length. The tabu list is referred to as *short term* memory.

Tabu search algorithms can be enhanced a great deal by using longer term memory. This is a big topic, but very briefly the idea behind longer term memory is to accomplish *intensification* and *diversification*. Intensification involves attempts to visit regions of the search space that seem promising. For example in a binary problem, a table can be kept of the frequencies of zeros and ones in each bit position that have been observed in solutions that have an objective value below some threshold. These frequencies can then be used to construct a solution that can be used to restart the search if a new best solution has not been found for a long time. Diversification involves attempts to visit regions that have not been visited. A frequency table can also be kept for all solutions visited and it can be used to construct solutions that are unlike the typical solutions encountered during the search. Therefore, appropriate information about solutions visited may be collected through a function LTMem(ν).

8.4.7 Genetic and Evolutionary Algorithms

A reasonable starting point for construction of a genetic algorithm (GA) would seem to be a "canonical GA." In order to discuss GAs as they were originally presented, it is useful to introduce the term *bit string*, which refers to a vector whose elements can be either zero or one. This is easily extended, but it is good place to start. For use, this is consistent with our example where the vector ν consists of the zero-one (or "bit valued") decision variables.

In contrast to the search methods discussed so far, a GA examines more than one solutions at once. This set of solutions is called population. That is, a simple GA begins with a population, i.e., a set, say POP, of many bit strings (each string of length n) and then recursively forms replacement populations by use of *crossover* and *mutation* operators. The simplest form of crossover is *one-point* crossover using two strings, a and b, as *parents* to produce two *children*, c and d. One of the parents is chosen using *fitness based selection*. That is, at random, but with a probability that decreases with increasing objective function value. The other parent is selected at random. A crossover point k is selected at random from integers between one and $n - 1$. Assign values to c and d as follows:

$$c_i \leftarrow \begin{cases} a_i & 1 \leq i \leq k \\ b_i & k < i \leq n \end{cases} \quad \text{and} \quad d_i \leftarrow \begin{cases} b_i & 1 \leq i \leq k \\ a_i & k < i \leq n \end{cases}$$

The mutation operator considers each bit position of a string and with some (low) probability changes the bit's value. The basic idea is that desirable bits will be retained in subsequent generations due to fitness based selection for crossover and a variety of new bit patterns are introduced due to random selection of the crossover point and due to mutation.

A simple genetic algorithm is shown in Figure 8.5. Table 8.3 describes the parameters and variables used by the algorithm. The functions FRSelect, Cross, and SwapMutate must deviate slightly from a truly "canonical" GA.

The function FRSelect (Fitness Ratio Select) returns the index into the population of a member selected at random with a probability of selection for each member equal to the *fitness* of the member divided by the sum of the fitness of all members in the population. Generally, GA's are described for maximization problems and the fitness is defined in the simplest case as the evaluation function value. Since we are minimizing, we use classic GA technology and define the fitness of a member as the negative of the evaluation function value.

Name	Description
PopSize	Population size (Parameter)
Pc	Crossover probability (Parameter)
Pm	Mutation probability (Parameter)
G_t	Population at iteration t; a vector of bit strings
i, j	Indexes into G_t of the selected member and mate, respectively
k	Crossover "point"

Table 8.3. Canonical GA: Parameters and Variables

```
randomly initialize population, G_0; t ← 0
REPEAT
   REPEAT
      FOR 1 ≤ i ≤ S BestCheck(G_t[i])
      i ← FRSelect(G_t)                                   (selection)
      IF URan(0,1) < Pc                                   (crossover?)
         j ← IRan(1, PopSize)                             (mate)
         k ← IRan(1, n)                                   (crossover point)
         place Cross(i, j, k) and Cross(j, i, k) in G_t+1
      ELSE                                                (no crossover)
         place string G_t[i] in G_t+1
   UNTIL G_t+1 has PopSize vectors
   FOR each 1 ≤ i ≤ PopSize                               (each Pop member)
      IF (URan(0,1) < Pm) SwapMutate(G_t+1[i])
   t ← t + 1
UNTIL TimeLimit() OR Convergence()
```

Fig. 8.5. Canonical Genetic Algorithm

The function Cross(i, j, k) can make use of a variety of alternative crossover schemes. The classic crossover mechanism is referred to as *one-point* crossover. This is intended to mimic chromosomal crossover in sexual reproduction. Each of the two parent strings is cut into two pieces with the leftmost k values in one piece and the rest in the other. The left part from the first string is combined with the right part from the second string to form the "child." Another alternative is *uniform crossover*, which constructs a child one element at a time. If both parents have the same value for an element, the child acquires that value. If it differs, then the value from one of the parents is selected at random (perhaps with greater probability assigned to the parent that corresponds to the better objective function value). Crossover for sequencing problems is also possible, but requires more notation than we want to develop for an overview of GA.

In a canonical GA, after new solutions have been created by crossover operations, a *mutation* operator is applied. This operator randomly perturbs the solution. Although it is not always described this way, it is very common for the mutation operator to be a few random moves using some simple neighborhood. The idea is to mimic nature and also to diversify the search. Figure 8.5 contains the function SwapMutate which performs a random swap move if the random number is less than a parameter, PM.

Regardless of which of the many crossover and mutation operators are used, it is extremely likely that a child solution will be infeasible. This is particularly true for complex scheduling problems with numerous, complicated constraints. There are two common ways to address this issue. One is to perform a *repair* to make changes to the solutions that will result in a feasible solution that resembles the result of crossover and mutation. Another possibility is to maintain a population not of solutions, but of *coded solutions* that imply feasible solutions as a result of *decoding* (perhaps by implying the solution of an LP or some other solvable problem). Crossover and mutation are applied to the coded solutions and the corresponding objective function values are obtained by decoding. The details depend on the particular problem being solved and are beyond the scope of our brief introduction.

GAs terminate when the population contains only solutions that are all the same or nearly all the same. The function labeled Convergence() contains the tests (typically based on parameters that are not shown). Mutation and random parent selection helps avoid early convergence.

There are many modern components of GAs that are not shown in the classic, canonical GA. Rather than using *generational replacement*, where the entire population is replaced by the results of crossover and mutation on the previous population, most modern GAs use *steady-state replacement* where offspring are either added to the existing population or discarded if it is a duplicate or has an objective function value that is not good enough. Often, a descent is performed on offspring and the result of the descent is added (perhaps after mutation). The selection of population members to be used in crossover is typically also much more complicated and depends more on the quality of the solutions or their anticipated effect on the diversity of the population.

Our goal in providing a brief description of SA, TS, and GA has been to provide enough background information to enable a modeler to participate in creating a working algorithm. Heuristic search is an active area of research. New ways of enhancing these three methods as well as new methods are being invented all of the time. A new method that is similar to heuristic search but comes from a different perspective and offers enhancement to optimization algorithms of all types is described in the next section.

8.5 Constraint Programming

The term *constraint programming* (CP) refers to a collection of modeling and solution methods. As the name implies, a cornerstone is the ability to model logical relationships directly. This goes far beyond the capabilities provided by SOS1 and SOS2 to allow a full set of logical and non-linear operators. For example, we can constrain production so that if SKU A is made then B cannot be made unless D is made and C is not made, etc. In addition, sequencing and precedence constraints can be modeled directly.

Another cornerstone of CP is *constraint propagation*, which facilitates *domain reduction*. The idea behind constraint propagation is that if a variable is fixed at some value, or has its range of possible values reduced, then this will have implications for the possible values of other variables due to interactions within the constraints. For example, suppose we have the following constraints:

$$
\begin{aligned}
&\dots \\
\delta_{1,1} + \delta_{2,1} &\le 1 \\
\delta_{1,1} + \delta_{3,1} &\ge 1 \\
&\dots \\
\delta_{i,t} \qquad &\in \{0,1\} \qquad i = 1,\dots,P,\ t = 1,\dots,T
\end{aligned}
$$

If the value of $\delta_{2,1}$ were to be fixed at one by a search algorithm, we could conclude that $\delta_{1,1}$ would have to be zero because of the first constraint shown. Since $\delta_{1,1}$ appears in the second constraint, this fact is propagated to it and used to reduce the domain of $\delta_{3,1}$ to be exactly one value, which is 1. So if $\delta_{2,1}$ is fixed at 1, we can fix the values of two more variables and thereby have a smaller problem to solve. Of course, these three variables may appear in more constraints and allow further reductions. CP solvers maintain a network representation of the constraints so that if a variable is fixed, the implications for the values of other variables can be quickly determined.

Although there are graphical user interfaces for various problem types, a majority of CP applications to date are done in conjunction with an object oriented programming language. Commercial CP packages provide extremely well developed base classes and operators. The range of problems that can be addressed using CP is somewhat narrower than the range offered by general heuristic search methods, so it is possible to develop more sophisticated interfaces. The generality, flexibility and power of CP interfaces is constantly improving.

Constraint propagation and domain reduction can be embedded in algorithms that fix, and perhaps free the values of variables (or restrict and expand their domains). Originally, CP algorithms all made use of a tree search that is vaguely similar to branch and bound, but works by fixing and then freeing variables. The process of freeing variables during expansion of a tree is referred to as *backtracking* because the search goes back to previously visited nodes in the tree. This form of search was prefered because it allows recovery of previously generated constraint network information. Lately, researchers have been making use of CP technology to enhance branch and bound algorithms as well as heuristic search. The CP software propagates the effect of newly fixed variables. Such applications are often referred to as *cooperating solvers*.

The ability to model complicated requirements make CP well-suited for scheduling problems. During the solution of a planning problem, the question arises as to whether a feasible schedule can be found that will result in producing the quantities given by the plan. If none can be found, then the utilization data or the changeover time data (or both) can be changed (i.e., increased). Hence, CP methods can cooperate by solving embedded sub-problems such as described in §8.4.3; CP methods have a comparative advantage when the scheduling problems involve logical constraints.

9. Some Stochastic Extensions

Now that we have developed some practical models for production planning within a supply chain and have outlined solution methods, we pursue some topics that are very uncertain. The intention of this chapter is to address some issues that are not included in commercial supply chain planning software and are just barely being addressed in the research literature. As such, they are speculative research topics. The models and solution methods that we describe here may, or may not, ultimately be adopted. However, the modeling issues are critical and must be addressed.

We use the term *stochastics* to refer to things that are best modeled as random. Just as very few things are truly linear, very few things are truly random. However, randomness is often the best model available. Suppose person A flips a coin and then looks at the result but does not show person B. Person A will not use a stochastic model of the result of the coin flip, but that is the best that person B can do. Our concern here is with situations where a stochastic model is the best that we can do, which is the case in a great variety of situations.

Stochastic models can be appropriate for any of the data elements in the models that we have developed. For example, the utilization fraction, $U(i,k)$, cannot be known with certainty. If the uncertainty is significant, then a stochastic model is appropriate. Demand data are often particularly uncertain, so stochastic models of demand are often used.

When dealing with a stochastic situation, it is often advisable to establish policies that protect against uncertainties. For example, the idea of setting a policy based on safety stock was mentioned in §6.3. Such policies are used to trigger placement of an order for a particular item when its inventory level reaches a *reorder point*. The reorder point policy can be improved so that on average, a certain amount of safety stock will still remain when the order is fulfilled and the new inventory arrives. This policy guards against uncertainty in demand as well as uncertainty in the lead time required for replenishment.

Another stochastic issue that is often addressed by establishing a policy is *random yields*. Particularly in electronics and biotechnology, but in other industries as well, the quantity that is actually produced by a process is best modeled as depending on the quantity that is started and a great deal of randomness. In these situations policies are sometimes established to start enough products so as to have a reasonable probability of obtaining the desired quantity upon completion. Any extra production is placed in inventory to reduce the requirements for the next production run.

Policies are a very important concept for dealing with stochastics, but we will defer discussion of policies until Chapter 10. We will not address them here for two reasons. First, there is already a rich literature and reasonable software support for establishing policies in a stochastic production environment. Inventory and yield policies are covered in a number of production planning texts. Second, a more compelling reason is that policies are operational controls. While these controls can provide data for the planning process (as in §6.3), the process of establishing them is not part of the tactical planning that we have developed. We proceed to consider two important topics for planning: load dependent lead times and creation of plans that take into account the possibility of major events.

9.1 Lead Times and Congestion

When we introduced mrp in Chapter 3, we noted that a potentially serious problem is that mrp models assume that the lead time is property of the item or part. We captured this assumption with the notation $LT(i)$ for the lead time required to produce an order of part i. In §3.5, we noted that the time required to produce a part really depends both on the part being produced and the number of parts in the waiting line ahead of the order. It is usually a very good approximation to say that actual lead times are proportional to the length of the waiting lines.

There is no perfect way out of this problem because it is not possible to foresee the future. We cannot know that a machine failure will produce long waiting lines or that a work group will have a great day and complete all of their work for the week on Monday. The best we can hope to do is to model the uncertainty about the future in some useful way. We need some understanding of waiting lines and then we can try to modify our planning models to anticipate some of the effects of waiting lines.

In the academic literature, waiting lines are referred to as *queues* and theories about waiting lines are grouped under the rubric *queuing theory.* At first it might seem that using a French word to substitute for a perfectly good pair of English words is valuable only to pompous academics. Perhaps. However, we find the word queuing to be useful as a reminder that we are not studying waiting lines themselves, but abstract models of them for the purpose of adjusting our planning models.

It is useful to make a distinction between the lead time, which is used for planning and *flow time*, which is the actual time between start of a job (or, in some cases, issuance of the order for the job) and its completion. The flow time can reasonably be modeled as random in most production settings.

We use the lead time for planning, but then we get some flow time which is usually not the same.

9.1.1 The Issues

If work arrived at a perfectly constant rate and work was performed at a perfectly constant rate, there would be no queuing. But that is not the reality that we must live with. Work arrives in bursts and it gets done in spurts with interruptions for machine failures and a host of other events. We need an abstract model in order to come to grips with queuing.

To be consistent with the jargon of queuing theory, we will make use of the word *server* to denote resources that perform work. The rate at which they perform work will be called the *service rate.* The rate at which work arrives will be called the *arrival rate.* These words are useful because they emphasize that we want to view queuing in an abstract way. The servers could be an individual resource, a resource group, a production line, a factory or a supplier.

The ratio of the arrival rate and the service rate is referred to as the *utilization* and is often represented with the Greek letter ρ. The utilization defined in this way corresponds pretty well to standard notions. For example, if work arrives at an average rate of 6/hour and the server can complete an average of 10/hour, then the server will be utilized 60% of the time in the long run and will be idle (or at least not servicing the items of interest in this example) 40% of the time. Notice that we are careful to use the words "long run" and "average" because in the short run, the server can be busy much more or less than 60%, and arrivals and service can be much faster or slower than usual.

For a given service rate, increases in the arrival rate are exactly proportional to increases in the utilization. For most of the models that we are interested in during tactical production planning the capacity is fixed for each resource, but we are making decisions that will effect the average arrival rate. This is done both by setting the level of production and via selection of the appropriate routing.

There are many ways to analyze queuing for both the long run and the short run. This important topic is the subject of a large research literature to which a few pointers are provided in the references. For the kinds of queuing analysis needed for supply chain planning, computer simulation is very popular.

Many commercial software packages are available with graphical user interfaces that allow one to literally paint a picture of a production facility and then input information about the bill of materials, routings, capacities and the nature of the randomness that may be present. The simulation software then recreates the operation of the system with realizations for the random variables. This can be repeated numerous times with numerous realizations and the results can be analyzed statistically to study the performance of the system and the characteristics of the queues.

Regardless of the method of analysis, a critical point is that queuing time is a non-linear function of the utilization as shown in Figure 9.1. The axis labels are vague because these are representative curves that are roughly accurate over a broad range of conditions. It is also qualitatively correct for both short run queuing statistics as well as long run averages.

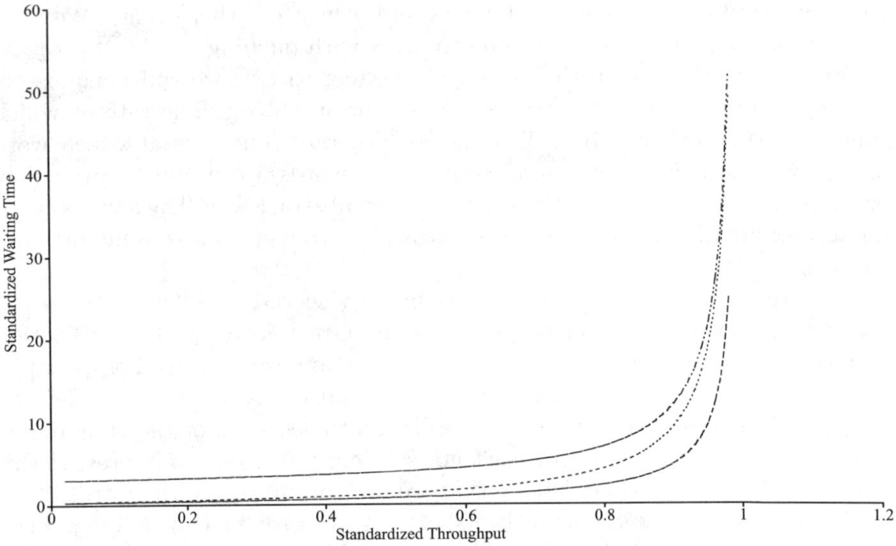

Fig. 9.1. Representative Curves Showing Queuing Versus Utilization for Three Different Situations

For fixed capacity planning, we can use the words "utilization" and "loading" interchangeably, where by loading we mean the amount of work assigned to a server (i.e., to a resource). Notice that in Figure 9.1 the sensitive part of the curve is at heavy loading. That is to say, given that the server is heavily loaded, fairly small changes in loading result in fairly large changes in queuing. This is in contrast with the left side of the curve where fairly large changes in the loading do not change the queuing very much.

It stands to reason that high average loadings result in long average queue lengths. If the server is very busy and anything goes wrong, work will pile up quickly. The shape of the curve depends on the characteristics of the arrival of work and the characteristics of the server. If the arrival process is fairly constant and the server is very reliable and consistent, then the curve will have a sharp bend very near a utilization of 1. In contrast, a server that is extremely failure prone and requires randomly varying times to complete work will begin to generate queuing with lower utilizations. In the worst case, such a server also faces a highly unpredictable pattern of work arrival. These three cases are depicted in Figure 9.1.

We can look at these curves another way. If one wants high utilization, one needs queues or (what is intended in supply chain management) synchronized arrivals. If work arrives in spurts then work is needed to keep the server busy during lulls in the arrival process. Either way we look at it, we can see that a reasonable definition for *a bottleneck* is a server that has high utilization and queues are building up. In a supply chain, we might refer to the server that has the highest average utilization as *the bottleneck*. If all of the other servers had significantly lower utilizations, we can call it a *sharp bottleneck*. This provides another view of the same concept that was developed in §6.6.

We can see immediately that if constant lead time estimates are to be used, they should be "tight" for SKUs that are not routed through a bottleneck and "loose" for those that are. Keeping the lead times near the processing time for non-bottlenecks will result in shorter overall lead times with no significant cost in terms of reduced throughput. Conversely, for SKUs that use a sharp bottleneck longer lead times are required both because that is realistic and because it can result in better bottleneck management.

It is realistic because longer queues are going to be present at bottlenecks and because we are planning and not scheduling. Schedule disruptions are often common and most disruptive at the bottleneck. Our planning models might as well acknowledge this. Providing the bottleneck with longer lead times allows more flexibility in scheduling it. This in turn is valuable because the throughput of a sharp bottleneck dictates the throughput of the larger organization.

A more sophisticated approach is to use *load dependent lead times*. This could be important in supply chain planning where decisions are being made regarding how to assign production to lines, factories, or subcontractors. In effect, these decisions determine the loading and therefore the flow times. The catch is that the best decisions also are determined by the flow time (or the lead time used for the optimization model).

9.1.2 Load Dependent Lead Times

Load dependent lead times are things that are not appropriately addressed by safety stock. The state-of-the-art planning technology for these things is to have a big meeting and argue. But we might also do it differently.

We attempt to extend the base SCP model as given in §5.5 to include selection of a reasonable lead time estimate based on the loading of the line. This is a research topic, but the model we develop here should help the reader grasp the issue of load dependent lead times and its effect on planning. The model we sketch here is very difficult to implement. We have implemented a small test version in order to give an example, but our implementation is not appropriate for practical applications. Our goal is to make the issues associated with load dependent lead times concrete by providing a model.

For SKUs effected by load dependent lead times we use a piecewise estimate of the utilization with enough breakpoints so that the difference in lead time from one breakpoint to the next is one time bucket. Needless to say, we do not want to go to the trouble of modeling variable lead times for every SKU. We only want to consider SKUs for which the lead time varies significantly as a result of decisions about how much to produce and how to route the production. Figure 9.1 illustrates that this is the same as saying "those SKUs that use resources that become heavily loaded." For such SKUs, we also do not necessarily want to model the lead time as a function of queuing at every resource along the routing. Typically, there will be only one or two *critical resources* that cause the lead time to vary with the load for a given SKU. So we have a slightly tautological definition of our data requirements. We need data concerning lead dependencies for those SKUs and those resources that are important to model; see Table 9.1. For ease of exposition we assume the existence of no more than one critical resource for each SKU.

$LT(i,0)$	Initial (and perhaps final) lead time for SKU i
$K(i)$	Critical resource for SKU i
$BP(i,r)$	A list indexed by r of utilization breakpoints for SKU i
$LT(i,r)$	The lead time that corresponds to $BP(i,r)$ for SKU i
$\rho'(i)$	Number of lead time breakpoints for SKU i
$H'(i)$	Holding cost for SKU i during the lead time

Table 9.1. Additional Data Required for Load Dependent Lead Times

Conceptually we can add load dependent lead times to a model by replacing $LT(i)$ by

$$\sum_{r=1}^{\rho'(i)} \rho_{i,r,t} LT(i,r) \tag{9.1}$$

to do the lead time selection for every SKU i effected by load dependent lead times. This expression uses a set of SOS1 (see §8.3) variables $\rho_{i,r,t}$, $r = 1,\ldots,\rho'(i)$ for each SKU i effected by load dependent lead times for every time t. The idea is that for each pair of i,t values, the value of ρ will be one for only one value of r. The effect will be that the corresponding LT value will have been selected.

Let r be an index into a discretized range of a macro $\hat{\rho}$, which is an approximation to the utilization of resource $K(i)$:

$$\hat{\rho}_{i,t}(x) \equiv \sum_{j=1}^{P} [U(j,K(i))x_{j,t} + S(j,K(i))\delta_{j,t}]$$

where $K(i)$ is the critical resource for SKU i. The summation over all SKUs using the index j is done because we are interested in setting the lead time breakpoint for SKU i based on the total utilization of its critical resource $K(i)$. We add the constraints for the binary variables $\rho_{i,r,t}$

$$\sum_{r=1}^{\rho'(i)} \rho_{i,r,t} = 1$$

for all SKUs i that are effected by load dependent lead times or else we require the sets to be SOS1 across r. Either way, the idea is that only one value of r will have a ρ value that is one for each product in each time period. This is the mechanism that is used to select the lead time that corresponds to the capacity utilizations. We also add the constraints

$$\sum_{r=1}^{\rho'(i)} BP(i,r)\rho_{i,r,t} \geq \hat{\rho}_{i,t}(x)$$

and finally

$$\sum_{r=1}^{\rho'(i)} \rho_{i,r,t}LT(i,r) \geq \sum_{r=1}^{\rho'(i)} \rho_{i,r,t+1}LT(i,r) - 1$$

for t between 1 and $T-1$ (one can add a constraint to make $\rho_{i,r,T+1}$ something reasonable). The last constraint keeps the lead time from jumping down by more than one time bucket. This is easy to see if the constraint is rewritten as

$$\sum_{r=1}^{\rho'(i)} \rho_{i,r,t+1}LT(i,r) - \sum_{r=1}^{\rho'(i)} \rho_{i,r,t}LT(i,r) \leq 1$$

If it were allowed to jump down by more than one, *passing* would be possible. That is, work started in a period would be modeled as being completed after work released in the next period; this clearly would not make sense in most practical cases. A closely related point is that jumps down of more than one would invalidate the summation from $i = LT(i,0)$ in the I macro.

To summarize, a base model such as **SCPa** (see §6.1) is modified to make use of the macro $\hat{\rho}$. To illustrate without a lot of clutter, we use a version of the objective function without the V and O terms. Such a new objective function would be to minimize

$$\sum_{t=1}^{T}\sum_{i=1}^{P}\left(A(i)I^{-}_{i,t} + H(i)I^{+}_{i,t} + H'(i)\sum_{r=1}^{\rho'(i)} \rho_{i,r,t}LT(i,r)x_{i,t} + C(i)\delta_{i,t}\right)$$

and the constraints

$$\sum_{r=1}^{\rho'(i)} \rho_{i,r,t} = 1 \qquad \forall\, i,t$$

$$\sum_{r=1}^{\rho'(i)} BP(i,r)\rho_{i,r,t} \geq \hat{\rho}_{i,t}(x) \qquad \forall\, i,t$$

$$\sum_{r=1}^{\rho'(i)} \rho_{i,r,t} LT(i,r) \geq \sum_{r=1}^{\rho'(i)} \rho_{i,r,t+1} LT(i,r) - 1 \qquad \forall\, i,t$$

$$\rho_{i,r,t} \quad \text{SOS1 across } r \qquad \forall\, i,t$$

are added. Notice that $\forall\ t$ means $t = 1, \ldots, T$ except for the third set of constraints (the ρ constraints), where the termination is at $T - 1$.

Because the lead times are modeled as varying with our decisions, the cost of holding inventory during the lead time is no longer a sunk cost. We have added the data element H' to account for this. However, one would expect that $H'(i) \approx H(i)$ would be an adequate approximation in most circumstances.

If we extend model **SCPa** to include load dependent lead times, the resulting model is shown as **SCPL** in Figure 9.2. The shorthand notation $\forall$ and $\in$ is abused in the interest of brevity to mean "for all appropriate values of the indexes listed." For example, $\forall\ t, \ell \notin \mathcal{L}(i)$ means the constraints are repeated for $t = 1, \ldots, T$ and for values of ℓ in the set formed by removing from $1, \ldots, P$ the indexes that appear in the list of substitute sets $\mathcal{L}$. In some situations, $\forall\ t$ means $t = 1, \ldots, T - 1$ and in other situations it means $t = 1, \ldots, T$; the appropriate limit will generally be clear from the context.

Unfortunately, this model is non-linear because of the H' term in the objective function which has $LT(i,r)$ times x_{it}. Since $LT(i,r)$, by using $\rho_{i,r,t}$, depends on x in this model, this term is not linear.

9.1.3 Solver Issues

In addition to the problem of the non-linear objective function, an important detail remains. We said that "conceptually" one can replace $LT(i)$ with expression (9.1). If heuristic search solvers or constrained programming solvers are used, then one can actually make the replacement but not for pure MIP solvers. The trouble is that the inventory macro would have expression (9.1) as the terminating index for summation.

For exact MIP solvers based on branch and bound solutions to linear programming relaxations, one cannot have variables as subscripts. This can be remedied if for any master SKU at most one alternate has load dependent lead times. For every constraint involving the lead time for an SKU i effected by load dependent lead times, $\rho'(i)$ constraints must be substituted, one for each possible value of r, and then the SOS1 variables ρ are used to make it so that only one of the constraints is meaningful.

Minimize:

$$\sum_{t=1}^{T}\left[\sum_{i=1}^{P}\left(A(i)I^{-}_{i,t}+H(i)I^{+}_{i,t}+H'(i)\sum_{r=1}^{\rho'(i)}\rho_{i,r,t}LT(i,r)x_{i,t}+C(i)\delta_{i,t}+V(i)x_{i,t}\right)\right.$$
$$\left.+\sum_{k=1}^{K}O(k,t)y_{k,t}\right] \qquad \textbf{(SCPL)}$$

subject to:

$$\begin{array}{lll}
I_{i,t}(x,\delta)+\sum_{\tau=1}^{t}D(i,\tau) & \geq 0 & \forall\, i,t \\
\sum_{i=1}^{P}[U(i,k)x_{i,t}+S(i,k)\delta_{i,t}] & \leq 1+y_{k,t} & \forall\, k,t \\
y_{k,t} & \leq F(k,t) & \forall\, k,t \\
\delta_{i,t} & \geq \frac{x_{i,t}}{M} & \forall\, i,t \\
\delta_{i,t} & \in \{0,1\} & \forall\, i,t \\
y_{k,t} & \geq 0 & \forall\, k,t \\
\sum_{r=1}^{\rho'(i)}\rho_{i,r,t} & =1 & \forall\, i,t \\
\sum_{r=1}^{\rho'(i)}BP(i,r)\rho_{i,r,t} & \geq \hat{\rho}_{i,t}(x) & \forall\, i,t \\
\sum_{r=1}^{\rho'(i)}\rho_{i,r,t}LT(i,r) & \geq \sum_{r=1}^{\rho'(i)}\rho_{i,r,t+1}LT(i,r)-1 & \forall\, i,t \\
\rho_{i,r,t} & \text{SOS1 across } r & \forall\, i,t \\
x_{\ell,t} & \geq 0 & \forall\, t,\ell\in\bigcup_{i=1}^{P}\mathcal{L}(i) \\
x_{\ell,t} & =0 & \forall\, t,\ell\notin\bigcup_{i=1}^{P}\mathcal{L}(i) \\
I^{+}_{\ell,t} & =0 & \forall\, t,\ell\in\bigcup_{i=1}^{P}\mathcal{L}(i) \\
I^{-}_{\ell,t} & =0 & \forall\, t,\ell\in\bigcup_{i=1}^{P}\mathcal{L}(i) \\
I^{+}_{\ell,t} & \geq 0 & \forall\, t,\ell\notin\bigcup_{i=1}^{P}\mathcal{L}(i) \\
I^{-}_{\ell,t} & \geq 0 & \forall\, t,\ell\notin\bigcup_{i=1}^{P}\mathcal{L}(i) \\
I^{+}_{i,t}-I^{-}_{i,t} & =I_{i,t}(x,\delta) & \forall\, t,\ell\notin\bigcup_{i=1}^{P}\mathcal{L}(i)
\end{array}$$

Fig. 9.2. SCPL Model

The implementation of this row selection mechanism is complicated. It is hard to write the macro for master SKUs in general because one needs to do row selection for the selected subset (i.e., in general one has to do two selections: one for the substitute and the other for the lead time. This can quickly become unwieldy). As stated, if there is only one substitute for which the lead times vary, then the implementation can be manageable, although still not trivial.

We are now back to the issue of defining the data. The goal here is to create a model that does a reasonable job of anticipating lead times that will result from load dependencies. Clearly, the maximum number of breakpoints needed is enough to make the lead time differ by one bucket for each breakpoint. The values for the breakpoints can be set using simulations, statistical analysis of historical data, or engineering estimates.

Note that values greater than 1 for $\hat{\rho}$ are possible so that LT can be given special values for overtime. It might go down again for overtime if that is realistic and then go back up if overtime reaches its capacity. Also note that zero is a legitimate possibility, especially for light loadings.

9.1.4 Example

To illustrate the notation, we extend the example developed in §3.1 and §4.1 which we review briefly. Part AJ8172, or SKU 1, is routed through resource 1 (HR-101) and resource 2 (MT-402). Part LQ8811, or SKU 2, is routed only through resource 1. For this example, we will assume that resource 1 is the bottleneck and that there is usually very little queuing at resource 2. The other SKUs are ordered from vendors and for this example we will assume that the vendor's overall utilization, and therefore the vendor's lead times, are not effected very much by the decisions that we make.

In the example developed to illustrate the mechanics of mrp and MRP II, we assumed that $LT(1)$ was 2 days and $LT(2)$ was 3 days. Suppose instead that based on analysis of past performance or a simulation study, the lead time really depends on the utilization of resource 1 with the lead time changing approximately at utilizations of 0.7, 0.9 and 1. The "default" lead times of 2 and 3 are valid between 0.7 and 0.9 and overtime assumed to be sufficient to reduce the lead time back down to default levels. So, to use our notation, $\rho'(i) = 4$ and $K(i) = \{1\}$ for $i = 1, 2$. The data for the lead time model is summarized in Table 9.2. The value for ∞ can be anything that is big enough to catch all possible values of $\hat{\rho}$. A value of 2 or 3 should be more than big enough and in many cases 1.5 might be adequate.

For the demand vector (20, 30, 10, 20, 30, 20, 30, 40) we solved model **SCPL** using a zero demand last period to handle boundary conditions and a crude approximation to the work in process holding cost term (H'). The solution can be found in a few seconds on a personal computer. The solutions are dominated by beginning and ending effects because we have used a simple model for illustration purposes. The main point is the large qualitative difference in the middle period of the optimal solutions for models with and without load dependent lead times.

r	i	$BP(i,r)$	$LT(i,r)$
1	1	0.7	1
	2	0.7	2
2	1	0.9	2
	2	0.9	3
3	1	1	3
	2	1	4
4	1	∞	2
	2	∞	3

Table 9.2. Data for Load Dependent Lead Time Example

If the lead times are all fixed at their nominal values, the subcontractor was only used for 10 units of capacity to avoid tardiness. However, this schedule would not really work because AJ8172a was being produced at a level that resulted in 100% utilization of its critical resource. This would result in congestion, long lead times and failure to perform according to the plan. Not only would AJ8172a experience longer lead times than planned, so would its components that use the same critical resource. Hence, the plan might not prove to have been realistic and therefore of limited value. The better solution comes from a plan that captures the fact that lead times go up with the utilization.

If we removed AJ8172b from the data, we found that for the cost structure in use, the optimal solution was to avoid long lead times and simply incur the tardiness costs. A comparison of the objective function values with and without an alternative gives an estimate of the value of having an alternative available. In many situations there are fixed costs associated with the existence of alternatives such as spare capacity or even secondary suppliers. The use of such a model can give a very concrete estimate of the tactical value of such alternatives.

9.1.5 Complications and Discussion

This is a simple model, but it could be better than ignoring the fact that lead times depend on the loading. The model is most effective where there is a sharp bottleneck. It is most needed in supply chain planning environments where there are multiple routings.

This model is really too simple if there is more than one resource in K. Suppose there are two resources and one is loaded to capacity and the other has almost no load. In this case, $\hat{\rho}$ will be about 1/2 which will suggest very little queuing, when in fact the heavy load on one of the two resources will probably result in significant queuing. If two or three bottleneck resources are in a series, interactions between them result in complicated, non-linear functions. If the presence of multiple bottlenecks is a serious planning issue, perhaps a remedy is to use the simple model given here to get a starting point for computer simulations that can be used to search for a better solution.

As was the case with changeovers, we can get some managerial insights from the models themselves. In §4.4.5 we noted that in addition to the extra capacity that one gets when changeover times are reduced, one gets the benefit of a production system that is easier to manage as evidenced by the much simpler planning models that are available. We find the same situation here.

A facility with balanced capacity is harder to plan for than one with a sharp bottleneck. With perfect balance, every resource is the bottleneck. To put it another way, there can be value in the purchase of excess capacity for inexpensive resources because it provides the simultaneous benefits of reduced queuing (therefore, reduced lead times) and simpler, therefore better, planning. In similar fashion, a facility that is operated very near its peak capacity is harder to plan for than one that has some capacity cushion. This is, of course, a delicate tradeoff. Utilization of capacity lowers average costs, but at the expense of longer lead times and more planning and management difficulties.

Pull based production control systems such as Kanban and CONWIP provide benefits for planning systems by putting a cap on the work in process and therefore on the sizes of queues. This has the effect of reducing the dependency of the lead time on the loading and therefore allows simpler, yet more precise, planning models to be used. Input/output control is a planning concept designed to have the same effect by coupling the planning process with execution results. One constrains the planned releases in the next period into a resource to be no greater than the completed work in the previous period. In this way, the WIP in the resource cannot grow and therefore neither can the lead times in the long run. These sorts of control systems provide benefits for planning systems that are discussed in the references provided in §10.3.5.

The model we have presented is not really suitable for practical application. Our goal was to sketch one approach in order to illustrate the issues associated with planning when the lead times vary significantly with the planned production quantities.

9.2 Scenarios

We refer to models that have no stochastic elements as being *deterministic*. Up to this point, all models have been deterministic because all of the data is assumed known. Even though the load dependent lead times model was designed to accommodate queuing effects that are the result of stochastics, the model itself made use of deterministic data.

We mentioned that none of the data can really be known with certainty. For example, we have specified that the utilization fraction be given as data, $U(i,k)$, but the utilization fraction would be more accurately modeled as a function $U(i,k,\xi)$, where ξ is a random variable. The same is true of demand and the other problem data. Whether it is worth the extra trouble to specify such a function depends on how important the uncertainty is.

In general, dealing with uncertainty is an important part of the job of production planners in a supply chain. Planners might express this based on the need to be concerned about "what-if." Such concerns are typically not expressed as functions of random variables, but rather as concerns about particular changes in the model data. For example, there might be a possibility of a new machine being purchased or the possibility of a new, large order being received. Hence, there really is not one set of data, but neither are there a set of complicated functions that indicate the random nature of the data. We represent this idea in an abstract way using a list of scenarios where each scenario is a complete set of data for the planning model.

9.2.1 The Issues

For tactical planning purposes, we are not interested in scenarios that capture a large number of relatively minor, high frequency events. We want to capture the few most important events. We are interested in cancellation of large orders, placement of large orders by a new customer, major disruptions at suppliers, competitors, or customers, etc. We might want to have contingency plans or we might want to hedge against these events.

Contingency plans are created for the purpose of being used once it becomes clear which scenario is occuring. For example, a plan might be developed for production once a large order is received. Hedging refers to the creation of plans that take action in advance of knowing what will happen. In the case of a large potential order, a hedged plan might call for increased production of some components even before the order is received.

We define the scenarios so that each one has a full set of data for the model. As a consequence most of the data in each scenario is the same as for all of the others and only major data elements differ. Some solution techniques and data management software can take advantage of this, but to keep our descriptions simple, we will subscript all data by the scenario number. We index the scenario set, $\mathcal{S}$, by s. Let the number of scenarios be given by S. For example, we give a capacity utilization estimate for each scenario and indicate the capacity utilization estimate for SKU i on resource k under scenario s as $U(i,k,s)$. Using this notation we can see that the scenario list is a simplified model intended to capture the important stochastic elements that would be present in the more complicated function $U(i,k,\xi)$.

For example, when we first developed MRP II in §4.1, we used capacity estimates as shown in Table 4.1 and reproduced here as Table 9.3, which is labeled as Scenario 1. Suppose that there is a chance that the capacity of

MT-402 would be doubled. This would create Scenario 2 with utilizations as shown in Table 9.4.

	Fraction Used By	
Resource	AJ8172	LQ8811
HR-101	1/480	1/960
MT-402		1/300

Table 9.3. Fraction of Capacity Utilized to Make Each SKU at Each Resource Under Scenario 1

	Fraction Used By	
Resource	AJ8172	LQ8811
HR-101	1/480	1/960
MT-402		1/600

Table 9.4. Fraction of Capacity Utilized to Make Each SKU at Each Resource Under Scenario 2

If we tried to implement the mrp plan the capacity utilizations would be as shown in Table 4.2, which is reproduced here as Table 9.5. Under Scenario 1, the mrp plan is not feasible because the capacity of MT-402 is exceeded in Period 3 (the utilization is computed to be 4/3, which is greater than 1). However, under Scenario 2 there would be sufficient capacity as shown in Table 9.5. The question is: How much LQ8811 should we plan to make in Period 3? Our decision will have implications both for the planned production of AJ8172 in subsequent periods as well as plans for orders for, and inventory position of, NN1100 and WN7342.

Period	Resource	Utilization under Scenario 1	Utilization under Scenario 2
3	HR-101	$\frac{100}{480} + \frac{400}{960} = \frac{5}{8}$	$\frac{100}{480} + \frac{400}{960} = \frac{5}{8}$
	MT-402	$\frac{400}{300} = \frac{4}{3}$	$\frac{400}{600} = \frac{2}{3}$
6	HR-101	$\frac{100}{480} = \frac{5}{24}$	$\frac{100}{480} = \frac{5}{24}$

Table 9.5. Anticipated Capacity Utilization for MRP II Example

We might want to just solve the problem for every scenario and look at the result. So, for example, we would formulate S different **SCPc** models and solve each one. We would then look at the solutions and use the information to help guide the solution that we ultimately use for planning purposes. It can be a little difficult to look through even a handful of large solutions and then make decisions.

This type of model management is supported by most modeling language packages as well as by the model management system that is part of the solver in Microsoft Excel. Each scenario is a separate model. The models and the solutions are maintained and indexed (perhaps using index names that identify the scenario) for retrieval. This sort of system is particularly useful for contingency planning.

A more sophisticated hedging model takes into account the likelihood of each scenario and suggests a solution that is in some sense a compromise between the solutions that are obtained using the data for each scenario considered separately. We refer to the probability of occurrence of s (or, more accurately, a realization "near" scenario s) as $Pr(s)$. Estimating the probabilities can be difficult. But it must be done either implicitly, as is the case when the scenarios are solved separately and the decision maker must weigh the various solutions, or explicitly. It seems better to be explicit about the assumptions that are used for planning and perhaps better to get the help of a computer to weight the various scenarios.

There are a lot of possible methods and models that consider the probabilities of each scenario. For financial planning models, one often wants to consider the tradeoffs between risk and return. In other situations, one might want to maximize the probability of achieving some threshold level of the objective function or minimizing the probability of a really bad objective value. But we argue that for tactical production planning, an appropriate objective is to consider the expected (or "average") value. One should minimize the expected cost or maximize the expected profit.

We offer two reasons for using expectations. First, production planning decisions are made many times per year so that the law of large numbers suggests that over even a fairly short time horizon, we should see roughly the average performance. This is especially true for production planning where the individual decisions are not usually of potentially catastrophic size (unlike, say, strategic planning). The second reason is that many of our models rely on costs that are best described as proxies. These proxy costs, such as the cost of tardiness, are very useful in helping to construct a good production plan. However, they make any attempts to set objective function thresholds or to trade off objective function values and risks dubious.

A third reason is that there is a possible solution technique available for expected value models. In fact, for two time period models there are a number of solution techniques available. These two-stage models can be appropriate for capacity planning, but in most production planning situations more stages are needed for a realistic model.

For multiple time stage problems with integer variables, there are very few ways to look for solutions at this time. For the methods that are available, at most a handful of time stages can reasonably be solved using currently available technology, so time buckets must generally be combined. Furthermore, the problems must be fairly small, so typically an aggregated plan has to be considered (see §6.6). In the next section we detail the progressive hedging solution method and the corresponding stochastic model.

9.2.2 A Multi-stage Probabilistic Model With Recourse

The thing that makes multi-stage stochastic models difficult to solve is that the model includes the fact that decisions in the future can depend on which random events actually occur. With a deterministic model, we have assumed that we know the future, so theoretically there would be no reason for any changes to the plan. We are aware that this is not the case, but that is a simplifying assumption that is made in a deterministic model.

In a multi-stage stochastic model, we must plan for decisions that are *conditional* on the realizations of random variables. In other words, we have to make the period one decisions and stick with them, but we can state that our period two decisions will depend on which values that data took in the first time bucket. (We use the words "time bucket," "stage" and "period" interchangeably.)

When we use a deterministic model, we know that we will really solve a new problem after we find out what happens in period 1. No one in their right mind would solve a 52 week **MRPII** model and then blindly follow the plan for the next 52 weeks regardless of what happens. A planner might create a plan for 52 weeks, execute the week-one-plan and then create a new plan the following week, using new estimates of demand and new beginning inventory levels, etc. However, the deterministic models do not explicitly include in them the fact that plans will be adjusted later. Stochastic models with recourse do.

This sort of model requires a great deal of notation, so rather than continuing with an ERP model that itself contains a lot of notation, we switch to a canonical MIP representation. This also allows us to provide an overview of the progressive hedging solution method that is part of the motivation for the way we represent the models.

Stochastic Programming Notation. Assume that we have S scenarios and T stages. For each scenario, s, and each stage, t, we are given a row vector $c(s,t)$ of length $n(t)$, a $m(t) \times n(t)$ matrix $A(s,t)$ and a column vector $b(s,t)$ of length $m(t)$. Let $N(t)$ be the index set $1,\ldots,n(t)$ and $M(t)$ be the index set $1,\ldots,m(t)$. For notational convenience let $A(s)$ be $(A(s,1),\ldots,A(s,T))$ and let $b(s)$ be $(b(s,1),\ldots,b(s,T))$.

The decision variables are a set of $n(t)$ vectors; one vector for each scenario. Notice that the solution is allowed to depend on the scenario. Let $X(s)$ be $(x(s,1),\ldots,x(s,T))$. We will use X as shorthand for the entire solution system of x vectors (i.e., $(X = x(1,1),\ldots,x(S,T))$.

If we were prescient enough to know which scenario would be the realization (call it s) and therefore the values of the random variables, we would want to minimize

$$f_s(X(s)) \equiv \sum_{t=1}^{T} \sum_{i \in N(t)} [c_i(s,t)x_i(s,t)] \qquad (\mathbf{P}_s)$$

subject to

$$\begin{aligned} A(s)X(s) &\geq b(s) \\ x_i(s,t) &\in \{0,1\} \qquad i \in I(t),\ t = 1,\ldots,T \\ x_i(s,t) &\geq 0 \qquad i \in N(t) - I(t),\ t = 1,\ldots,T. \end{aligned}$$

where $I(t)$ defines the set of integer variables in each time stage. Discounting could easily be added. The notation AX is used to capture the usual sorts of single period and period linking constraints that one typically finds in multi-stage linear programming formulations such as **SCPa**.

Since we are not prescient, we must require solutions that do not require foreknowledge and that will be feasible no matter which scenario is realized. We refer to solution systems that satisfy constraints with probability one as *admissible*. We refer to a system of solution vectors as *implementable* if for scenario pairs s and s' that are indistinguishable up to time t, is true that $x_i(s,t') = x_i(s',t')$ for all $1 \leq t' \leq t$ and each i in each $N(t)$. We refer to the set of all such solution systems as $\mathcal{N}_{\mathcal{S}}$ for a given set of scenarios $\mathcal{S}$.

$$\min \sum_{s \in \mathcal{S}} [\Pr(s) f_s(X(s))] \qquad (\mathbf{P})$$

subject to

$$\begin{aligned} A(s)X(s) &\geq b(s) \qquad s \in S \\ x_i(s,t) &\in \{0,1\} \qquad i \in I(t),\ t = 1,\ldots,T,\ \ s \in S \\ x_i(s,t) &\geq 0 \qquad i \in N(t) - I(t),\ t = 1,\ldots,T,\ s \in S \\ X &\in \mathcal{N}_{\mathcal{S}} \end{aligned}$$

Unless time travel becomes possible, only solutions that are implementable are useful. Finding a solution for each scenario can be a useful exercise, but in the end one particular set of decisions must be implemented for the current period. It would be nice if the set of decisions happened to be optimal for the scenario that ultimately unfolds; however, that is extremely unlikely. Hence, a stochastic programming solution methodology must provide solutions that do not depend on knowing which scenario will occur. When working with multi-stage problems, this means that the solution at each stage can only depend on information that is available at that time stage.

Solutions that are not admissible, on the other hand, may have some value. Although some constraints may represent laws of physics others may be violated slightly without serious consequence. We would prefer an algorithm that produces admissible solutions, but might settle for a solution that violates some constraints, if the violations are not too severe.

The progressive hedging algorithm (PH) described below assures implementable solutions at all iterations and admissibility on convergence. If the algorithm has to be terminated before convergence, the solution at that point will have the critical property of implementability although it may or may not be admissible. The longer the algorithm runs, the more likely it is to have found a solution that has both desirable properties.

Approaches other than PH are possible for some types of integer problems. A straightforward solution method is based on a reformulation called the deterministic equivalent (DE). The DE is produced by reformulating the original problem so that dependence on scenarios is eliminated implicitly. This will be defined in a rigorous fashion after more notation has been developed; but for now we note that for mixed integer problems the resulting instance will typically be too large to be solved exactly even though smaller instances can be solved to optimality using the DE formulation.

We are interested in the more general case where there can be any number of integer variables in any and all stages and where any of the data can be stochastic. We are also interested in methods that can accommodate more than two stages and instances that are potentially too large for exact methods, even for one scenario. In order to have all of these features, we must abandon exact methods and make use of heuristics.

9.2.3 Progressive Hedging

The progressive hedging algorithm as described here is both intuitively appealing and has desirable theoretical properties: it converges to a global optimum in the convex case, there is a linear convergence rate in the case of a linear stochastic problem, and if it converges in the non-convex case (and if the sub-problems are solved to local optimality) then it converges to a locally optimal solution.

To begin the PH implementation for **P**, we organize the scenarios and decision times into a tree. The leaves (i.e., terminals of the tree) correspond to scenario realizations. The leaves are grouped for connection to nodes at time T. Each leaf is connected to exactly one time T node and each of these nodes represents a unique realization up to time T. The time T nodes are connected to time $T-1$ nodes so that each scenario connected to the same node at time $T-1$ has the same realization up to time $T-1$. This is continued back to time 1 (i.e., "now"). Hence, two scenarios whose leaves are both connected to the same node at time t have the same realization up to time t. Clearly, then, in order for a solution to be implementable, it must be true that if two scenarios are connected to the same node at some time t the values of $x_i(t')$ must be the same under both scenarios for all i and for all $t' \leq t$.

To illustrate the notation developed thus far, we consider a very small example with three decision stages (so $T = 3$) and two decisions per period, one of which is binary the other of which is bounded above zero and only three additional constraints. For any given scenario s the problem $\mathbf{P}_s$ is to minimize:

$$\sum_{i=1}^{3}\sum_{j=1}^{2} c_j(s,i)x_j(s,i)$$

subject to

$$\begin{aligned}
\sum_{j=1}^{2} a_{1j}(s,1)x_j(s,1) &\leq b_1(s,1)\\
\sum_{j=1}^{2} a_{2j}(s,1)x_j(s,1) &\leq b_2(s,1)\\
\sum_{i=1}^{3}\sum_{j=1}^{2} a_{1j}(s,i)x_j(s,i) &\leq b_1(s,3)\\
x_1(s,t) &\in \{0,1\} \qquad t = 1,2,3\\
x_2(s,t) &\geq 0 \qquad t = 1,2,3
\end{aligned}$$

Suppose that all data for $t = 1$ are known and the data for $t > 1$ will become available before we will have to commit to the decision variables for these times. Further suppose that the only data that are stochastic for $t > 1$ are c_2 and a_{12}. Now suppose that at $t = 2$, we will have $c_2 = 2$ and $a_{12} = 3$ with probability 0.6 and $c_2 = 3$, $a_{12} = 3.7$ as the only other possibility. If the first case occurs, then we will know with certainty that for $t = 3$, the values of c_2 and a_{12} will both be 6, but in the second case there is a 0.5 chance that they will be unchanged from $t = 2$ values and a 0.5 chance that they will both be 6. This is seen much more easily with a scenario tree graph as shown in Figure 9.3. This graph uses a circle for both nodes and leaves. For each circle, this graph shows the pair (c_2, a_{12}) along with the unconditional probability of realizing the scenario(s).

If we are not interested in contingency planning, the solution for any particular scenario may not be of any value to us. Ultimately, we want to be able to solve the problem of minimizing the expected value of the objective function subject to meeting the constraints for all of the scenarios and also subject to the constraint that the solution system will be implementable. The way we obtain such solutions is to use progressive hedging. For each scenario s, approximate solutions are obtained for the problem of minimizing, subject to the constraints, the deterministic f_s plus terms that penalize lack of implementability. They make use of a system of row vectors, w, that have the same dimension as the column vector system X, so we use the same

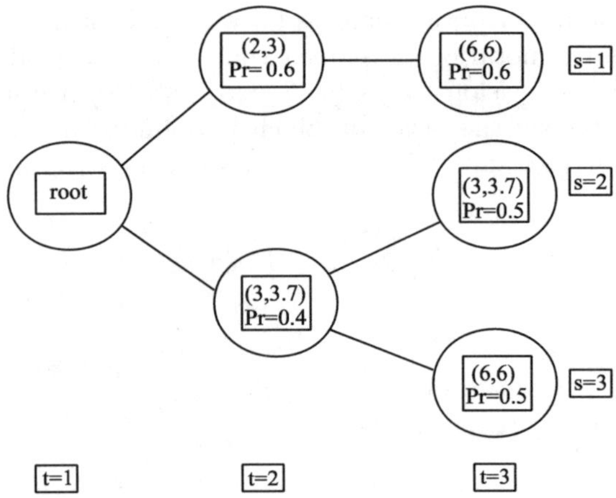

Fig. 9.3. Scenario Tree for the Small Example Problem

shorthand notation. For example, $w(s)$ means $(w(s,1),\ldots,w(s,T))$ for the multiplier system.

In order to give a formal algorithm statement, we need to formalize some of the scenario tree concepts. We use $\Pr(\mathcal{A})$ to denote the sum of $\Pr(s)$ over all s for scenarios emanating from node $\mathcal{A}$ (i.e., those s that are the leaves of the sub-tree having $\mathcal{A}$ as a root also referred to as $s \in \mathcal{A}$). We use $t(\mathcal{A})$ to indicate the time index for node $\mathcal{A}$ (i.e., node $\mathcal{A}$ corresponds to time t).

Let the operator $\overset{HS(x')}{\longleftarrow} x$ mean "assign the result of a heuristic search optimization that attempts to find the argument x that minimizes the right hand side of the expression with the search beginning at x'." This is described in detail in §8.4. We use $\overset{HS(\bullet)}{\longleftarrow} x$ when the PH algorithm does not suggest a starting solution for the search.

For typesetting convenience, we consider the constraints for problem $\mathbf{P}_s$ to be represented by the symbols "$X(s) \in \Omega_s$." We use $X(t;\mathcal{A})$ on the left hand side of a statement to indicate assignment to the vector $(x_1(s,t),\ldots,x_{N(t)}(s,t))$ for each $s \in \mathcal{A}$. We refer to vectors at each iteration using a superscript; e.g., $w^{(0)}(s)$ is the multiplier vector for scenario s at PH iteration zero. The PH iteration counter is k.

If we defer briefly the discussion of termination criteria, a formal version of the algorithm (with step numbering that matches in the informal statement just given) can be stated as follows taking $r > 0$ as a parameter.

1. $k \leftarrow 0$
2. For all scenario indexes, $s \in \mathcal{S}$

$$X^{(0)}(s) \overset{HS(\bullet)}{\longleftarrow} X(s) f_s(X(s)) : X(s) \in \Omega_s$$

and

$$w^{(0)}(s) \leftarrow 0$$

3. $k \leftarrow k+1$
4. For each node, $\mathcal{A}$, in the scenario tree, and for $t = t(\mathcal{A})$

$$\overline{X}^{(k-1)}(t; \mathcal{A}) \leftarrow \sum_{s \in A} \Pr(s) X(t; s)^{(k-1)} / \Pr(\mathcal{A})$$

5. For all scenario indexes, $s \in \mathcal{S}$

$$w^{(k)}(s) \leftarrow w^{(k-1)}(s) + (r) \left(X^{(k-1)}(s) - \overline{X}^{(k-1)}(s) \right)$$

and

$$\begin{aligned} X^k(s) \stackrel{HS(X^{(k-1)}(s))}{\longleftarrow} & X(s) f_s(X(s)) + w^{(k)}(s) X(s) \\ & + r/2 \left\| X(s) - \overline{X}^{k-1}(s) \right\|^2 \\ & : X(s) \in \Omega_s . \end{aligned}$$

6. If the termination criteria are not met, then go to Step 3.

The termination criteria are based mainly on convergence, but we must also terminate based on time because non-convergence is a possibility. Iterations are continued until k reaches some pre-determined limit K or the algorithm has *converged* which we take to mean $\overline{X}^k(s)$ is *sufficiently close* to $\overline{X}^{k-1}(s)$ for all s. One possible definition of "sufficiently close" is to require the distance (e.g., Euclidean, indicated by $\|\cdot\|^2$) to be less than some parameter. A much better choice is to consider only integer components of $\overline{X}$ and require equality. This requires no metric or parameter and typically occurs after far fewer iterations. Once the integer components of a solution are (assumed) known, the real valued variables are obtained by solving the deterministic equivalent problem (DE) with the integer values fixed. Although the DE is much larger than the original sub-problems, fixing all of the integers can more than mitigate this when there are even a modest number of integer variables. (It would also be possible to solve the problem once the integers have been fixed using any other method for multi-stage, linear stochastic optimization). The only potential drawback to convergence based only on integers (*integer convergence*) is that there is no theoretical guarantee that a better solution would not be obtained by waiting for convergence of the entire solution system (*full convergence*).

The vectors w can be interpreted as a dual price for the (implied) implementability constraints. In other words, they have an interpretation as a value of information. The algorithm is often classified as a "dual" method because we are essentially searching for good values of w. As we have described it, the direct search will be for X and the values of w are implied.

An overall interpretation of the algorithm is that it provides an optimal consensus solution. Hedging is a systematic way to find a compromise between the solutions for each scenario. The compromise takes into account the probabilities and effects of each scenario and furthermore the fact that future decisions can depend on the information that has become available over time.

9.2.4 A PH based Heuristic for SCPc

For large problems with integer variables, stochastic problems are often too large to solve to optimality. Hence, PH serves as the foundation for the creation of heuristic algorithms. These algorithms benefit from designs that take into account the special structure of the optimization problems to be solved.

The design of PH based heuristics for **SCPc** (and problems of that type) is aided by analysis of the binary variables. Although they can be viewed as indicators of production in a period, they can also be viewed as enabling variables. This view is useful for our purposes. When δ_{it} is one, there is a cost penalty in the objective function to cover changeover costs and there is potentially a reduction in the available capacity as a result of changeover times. When this cost and capacity penalty is incurred, production of SKU i in time t is enabled.

We exploit the importance of these binary variables to create an implementation of PH that avoids explicit inclusion of a quadratic term. The heuristic algorithm forces convergence only for the binary variables. In other words, in Step 5 of the PH algorithm, the w vector is updated only for the δ variables and the objective function is penalized only for the deviation of δ variables from their mean value at a node.

The objective function penalty for these binary variables is implemented by simply using the absolute value of the difference. This is sensible, since this is equal to the squared difference for binary variables. To linearize the absolute value we add variables and constraints to obtain the positive or negative difference, each of which itself has non-negative value. For values δ and $\overline{\delta}$, the variable p and m can be used if the following constraints are added

$$\begin{aligned} p - m &= \delta - \overline{\delta}, \\ p &\geq 0 \\ m &\geq 0 \end{aligned}$$

One of the things that makes the δ variables so important is their potentially large objective function coefficients. This raises some issues with respect to scaling. Rather than rescale the problem, we modify the update of the w vector to be scaled by half the δ objective function coefficient. In Step 5, the assignment of an updated w value is modified to be:

$$w^{(k)}(s) \leftarrow w^{(k-1)}(s) + (r)\left(\delta^{(k-1)}(s) - \delta^{(k-1)}(s)\right) C(i)/2$$

The penalty in the objective function is scaled in the same way.

For **SCPc** we extend the integer convergence criterion, which is that the PH algorithm is terminated after a certain number of integer variables have converged. The remaining variables are left free and the DE is solved to optimality. The idea is that it is the integers that cause the solution times to be excessive, so if enough of the integer variables can be fixed in the DE at their converged values, the DE is then tractable.

The central role of the production indicators in the planning model motivates a heuristic search at termination that we describe here. Rather than simply fixing those variables that have converged, we make use of ideas loosely motivated by GRASP. After κ iterations of PH, it is terminated. For each binary variable that has converged, its value is fixed in the DE with probability α. By varying α and κ the computational effort and expected solution quality can be controlled.

10. Research Directions and References

The material that we have provided in the previous chapters has its foundations in existing literature, which will be referenced and put into perspective in this chapter. This also includes some historical remarks which should allow the reader to follow the evolution of supply chain planning up to today.

Furthermore, the models and methods that we have described in the previous chapters are nested in a number of important areas of ongoing research. In this chapter we also give the reader some pointers to the research literature. This is not intended to be exhaustive, but rather to provide the reader with connections to the literature. An exhaustive review is out of question because the respective literatures are so large and they are evolving all the time. However, the references cited here should provide a more than adequate entry point for further access to these research areas.

10.1 Supply Chain Management

A good definition of a supply chain was provided by Ganeshan and Harrison (1995):

> A supply chain is a network of facilities and distribution options that performs the functions of procurement of materials, transformation of these materials into intermediate and finished products, and the distribution of these finished products to customers. Supply chains exist in both service and manufacturing organizations, although the complexity of the chain may vary greatly from industry to industry and firm to firm.

Here we give a brief discussion of those supply chain management issues that appear to have an influence on the overall management discussion when it comes to implementation of our models in practice. We start with a historical perspective on the evolution of logistics and end with a couple of thoughts concerning what might become important, especially for production planning within a supply chain.

Since our aim is not writing a general textbook on supply chain management we should mention that over the last couple of years a number of good textbooks have been written that deal with tactical and strategic supply chain management issues; see, e.g., Handfield and Nichols (1999), Shapiro (2001b), Bowersox et al. (2002), Simchi-Levi et al. (2002), Chopra and Meindl (2003). Consideration of the tactical and operational level, however, can be primarily found in academic journals or in edited books with collections of papers such as Tayur et al. (1999), Klose et al. (2002), de Kok and Graves (2003), Dyckhoff et al. (2004) or Stadtler and Kilger (2005).

10.1.1 The Evolution of Logistics

Before the words "Supply Chain Management" became popular, many of the activities associated with these words were referred to as *logistics* and others as production planning. While we consider production planning later, let us start with the primary objective of logistics. It can be described as the delivery of the right product in the right place at the right time at the least costs; see, e.g., Bowersox (1974), Christopher (1986). Historically, most organizations had considered logistics to be deserving of modest priority.

Prior to 1950, logistics was treated on a fragmentary and often secondary basis. Bowersox (1974) sees two major factors for this neglect and subsequent development. First, prior to the time that computers emerged and before applied analytical tools were generally at the disposal of business, there was no reason to believe that an integrated attack on logistical activities would accomplish improved performance. Second, the prolonged profit squeeze of the early 1950s created an environment conducive to the development of new cost control systems. Integrated logistics provided a productive arena for new methods of cost reduction.

In the period from the mid 1950s to the mid 1960s the concept of integrated logistics crystallized. According to Bowersox (1974) the economic climate at that times was responsible for the "flurry of attention to logistical problems." During the early 1960s, the horizons of emerging fields of integrated logistics began to expand. Emphasis began to shift towards a penetrating appraisal of the improved customer service capabilities enabled by a highly integrated logistics system.

The period of the 1970s was characterized by the integration of the intra-organizational logistics functions. Davis and Brown (1974) define *logistics management* "as the managerial responsibility of organizing, controlling, directing, staffing, and coordinating product flow from the point of initial procurement to the point of ultimate consumption." This definition encompasses the activities of purchasing, inventory control, material handling, site determination, warehousing, packaging, order processing, and transportation in a company. Furthermore, it should bridge the gap between the inbound flow of raw materials and the distribution of finished products. Also, Bowersox (1974) emphasized the need for an integrated treatment of intra-organizational functions (refer to the logistics management process depicted in Figure 10.1). He defines the *logistical mission* as the development of "a system that meets the stated corporate customer service at the lowest possible dollar expenditure." Development of a satisfactory program requires two levels of adjustment: integration of the logistical system with other corporate systems like the production system, the marketing system, or the finance system and development of total cost balance between logistical system components such as facilities, communication, inventory, transportation, and material movement.

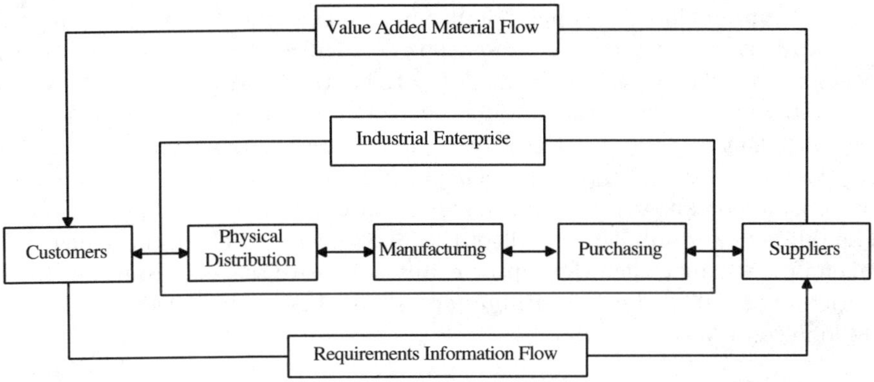

Fig. 10.1. Logistics Management Process from Bowersox (1974)

Since the early 1980s the integration of company-overlapping aspects in terms of logistics have become evident. In this connection the term *Supply Chain Management* was mentioned for the first time by Oliver and Webber (1982) (as noted by Christopher, 1999). Four aspects in which supply chain management differs significantly from classic materials and manufacturing control have been emphasized:

- Supply chain management views the supply chain as a single entity rather than relegating fragmented responsibility for various segments in the supply chain to functional areas.
- Supply chain management calls for strategic decision making. *Supply* is a shared objective of practically every function in the supply chain and is of particular strategic significance because of its impact on overall costs and market share.
- Supply chain management provides a different perspective on inventories which are used as a balancing mechanism of last, not first, resort.
- Supply chain management requires a new approach to systems: integration, not simply interfaces, is the key.

Over time the interest in inter-organizational integration has increased. According to Cooper et al. (1997) there is a definite need for the integration of business operations in the supply chain that goes beyond logistics. In addition to the internal functions of an organization (e.g., logistics, manufacturing, marketing, research) they see a need to integrate external organizations (e.g., consumer, suppliers, customer) in the product process in order to reduce the time-to-market on new product introductions. The integration of business processes across the supply chain that adds value for customers is what they are calling supply chain management. This definition is similar to the original definition of the term.

The references investigated so far imply that supply chain management is to some extent a new slogan and not a completely new concept. According to its original definition supply chain management means the integration of independent organizations. A closer look at Figure 10.2 reveals that integration has taken place since the middle of the 1950s (stage two). Of course, every stage was regarded as different *independent units* which have been integrated. But at some level of abstraction it does not matter whether only business functions (e.g., procurement, production, sales) or entire companies are considered. The underlying principle of integration is the same in all cases. The difference is only the extent of the *Supply Chain*. This aspect has to be taken into account when the supply chain is defined; see, e.g., Stevens (1989), Christopher (1999), Lee and Billington (1993), Lamming (1996), Larson and Halldorsson (2004).

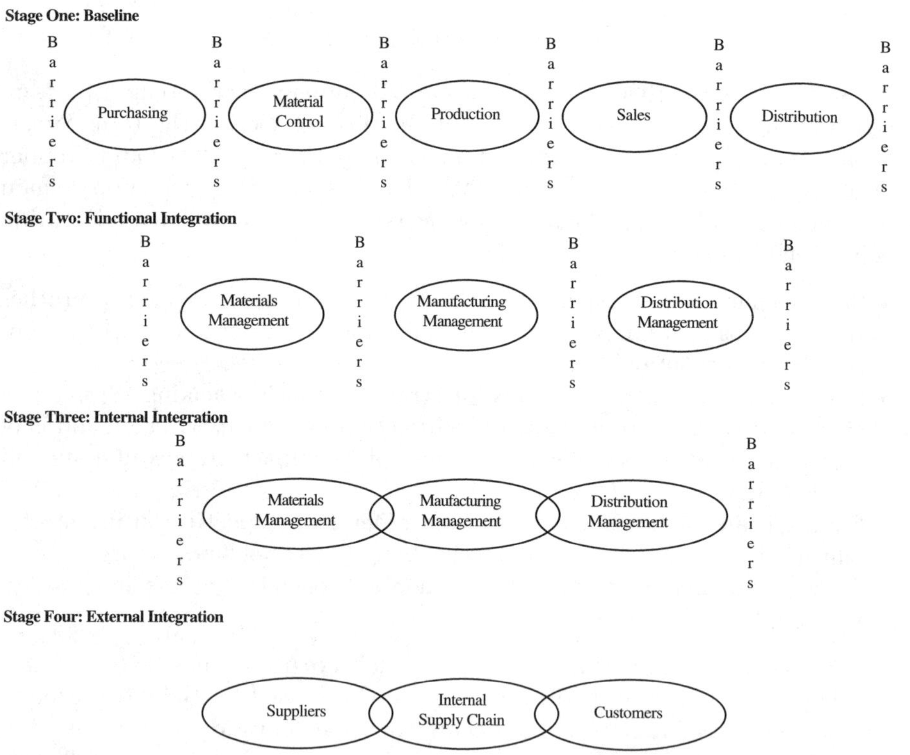

Fig. 10.2. Stages in the Evolution of Logistics from Stevens (1989)

10.1.2 Closed Loop Supply Chains and Reverse Logistics

In Europe manufacturers and importers of various products are legally obliged to take-back and recover their products after use. In response, manufacturers have set up collection and recycling networks eventually including a network of regional storage centers where products that are collected via municipalities and retailers are sorted and consolidated and then shipped to some recycling subcontractors. Two different and yet closely related fields are emerging in this area, reverse logistics and closed loop supply chain management.

Reverse logistics deals with returning waste materials and used products to the producer. While the functionality of logistics as we discussed it above is often referred to as forward logistics, the collection and recovery of used products refers to reverse logistics. Once forward and reverse logistics are combined in the sense of reusing recovered and used products for remanufacturing and delivering those remanufactured products into customer markets again, we speak about closed loop supply chains. We should mention that, in the same spirit as having somewhat loosely coupled supply networks and not necessarily just pure chains (see the definition on page 187), closed loop supply chains are more of a network including cycles than just pure chains.

Closed loop supply chains assume product returns which may imply remanufacturing as well as disposal. Having our lead time discussion in mind, we have a situation where remanufacturing lead time functions can be significantly different from production lead times. Nevertheless, from a modeling standpoint our models may be used as a starting point for developing extended and useful models for reverse logistics and closed loop supply chain management.

The importance of reverse logistics as well as remanufacturing used products into new ones has been widely recognized in the literature and in practice. Good sources on various aspects of closed loop supply chains and reverse logistics can be found, e.g., in Fleischmann et al. (2001), Fleischmann (2001) as well as some of the contributions in Guide Jr. and van Wassenhove (2003), Dyckhoff et al. (2004). Savaskan et al. (2004), e.g., consider the problem of choosing an appropriate reverse channel structure for the collection of used products from customers. Options taken into account are collection by the manufacturer himself, by an existing retailer or by subcontracting. An empirical study within the automotive aftermarket industry was undertaken by Richey et al. (2005) again underlining the importance of this growing field.

10.1.3 The Importance of Information Technology

As mentioned in §10.1.1 prior to the time that computers emerged and before applied analytical tools were generally at the disposal of business, the overall process logistics was treated on a fragmentary basis. Gains in computing speed, coupled with improvements in communication and the flexibility of data management software, have promoted a range of opportunities prevalent to supply chain management and supply chain planning. However, competitive advantage in supply chain management is gained not simply through faster and cheaper communication of data.

Shapiro (1999) points out that "to effectively apply information technology (IT) in managing its supply chain, a company must distinguish between the form and function of *Transactional IT* and *Analytical IT*." Transactional IT comprises acquiring, processing, and communicating raw data about a company's past and current supply chain operations, and the compilation and dissemination of reports summarizing these data. Analytical IT evaluates supply chain decisions based on models constructed from so-called supply chain databases. Usually these databases are derived from transactional data. Analytical IT is comprised of these supply chain decision databases, as well as modeling systems and communication networks linking corporate databases to them. It is concerned with analyzing decisions over short, medium, and long term futures.

According to Shapiro (1999, 2001a) inter-temporal coordination of supply chain decisions has received far less attention than functional coordination. Current efforts to improve supply chain management using IT and business process redesign have only focused on the operational and strategic levels with radically different time frames, planning concerns and organizational needs. Little effort has been made to link analytic tools and databases at the two extreme levels of planning.

Inter-temporal as well as functional integration can be achieved by the application of a suite of optimization modeling systems which take operational, tactical, and strategic aspects into account. These analytical IT systems are linked to overlapping supply chain databases created in large part from data provided by transactional IT systems. Figure 10.3 depicts a possible *Supply Chain System Hierarchy* comprised of optimization modeling systems and transactional systems responsible for inter-temporal and functional integration of supply chain activities in a manufacturing and distribution company with multiple plants and distribution centers.

As IT and supply chain management continue to improve and modeling applications expand, it is expected that more and more companies will implement versions of the entire system hierarchy in the near future. The subsystems of the hierarchy are:

- *Enterprise Resource Planning (ERP):* managing the company's transactional data on a continuous, real-time basis, i.e., standardizing data and information systems for order entry, financial accounting, purchasing, and many other functions, across multiple facilities and business units,
- *Materials Requirements Planning (mrp):* developing net requirements of raw materials and intermediate products to be manufactured or ordered from vendors to meet demand for finished products,

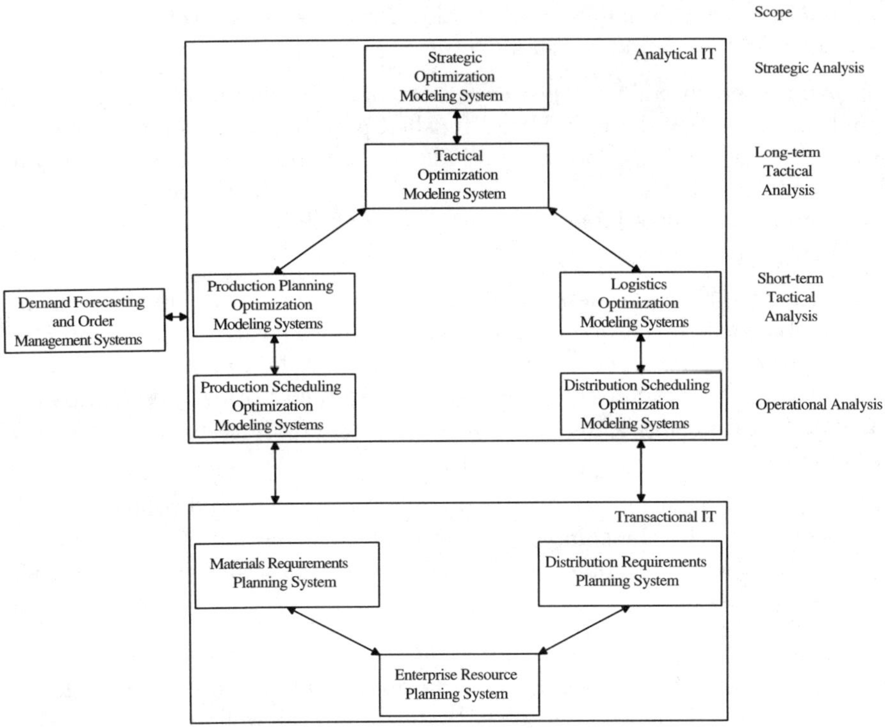

Fig. 10.3. Supply Chain System Hierarchy from Shapiro (1999)

- *Distribution Requirements Planning (DRP):* scheduling in-bound, inter-facility, and out-bound shipments through the company's logistics network, taking into account a wide range of transportation factors such as vehicle loading and routing, consolidation, modal choice, channel selection, and carrier selection,
- *Demand Forecasting and Order Management:* combining data about current orders with historical data to produce requirements for finished products to be met by operational, tactical, and strategic plans,
- *Production Scheduling:* addressing operational decisions at each plant of a supply chain, e.g., the sequencing of orders on a machine, the timing of major and minor changeovers, or the management of WIP,
- *Distribution Scheduling:* determining vehicle schedules and deciding on a short-term basis which distribution center should serve each market based on inventory availability,

- *Production Planning:* determining multi-period and multi-stage master production plans of manufacturing, along with resource levels and resource allocations, that minimize manufacturing costs,
- *Logistics:* determining a logistics master plan for the entire supply chain that analyzes how demand for all finished products in all markets will be met over the next appropriate period, and assigning markets to distribution centers and other facilities responsible for sourcing them with the goal of minimizing controllable transportation, handling, warehousing, and inventory costs across the entire logistics network,
- *Tactical Optimization:* determining an integrated supply / manufacturing / distribution / inventory plan for the company's entire supply chain over the appropriate couple of periods with respect to minimizing total supply chain costs of meeting fixed demand but also incorporating estimated demands based on forecasts, or to maximize net revenues if product mix is allowed to vary,
- *Strategic Optimization:* analyzing resource acquisition and other strategic decisions faced by the company such as the construction of new manufacturing facilities, development of acquisitions, or the design of supply chains for new products.

The application of any optimization modeling system in the system hierarchy requires inputs from a supply chain database that is created by transforming transactional data found in the ERP, mrp, DRP as well as the forecasting and order management system.

Some of the principles for creating and exploiting decision supply chain databases are as follows; see Shapiro (1999):

- *Adapt Managerial Accounting Principles in Computing Costs*
  For the purpose of decision making, the managerial accounting or modeling practitioner must develop relationships between direct and indirect costs, rather than point estimates of them.
- *Aggregation*
  As it is not necessary or desirable to describe operations at the individual SKU level for the purpose of strategic or tactical planning, the modeling of supply chain operations should incorporate suitable aggregation of products, customers, and suppliers.
- *Incorporation of External Data Concerning Suppliers, Markets, and Economies*
  Transactional data about the company's operations are not sufficient in scope for supply chain analysis of a strategic and tactical nature. An optimization model may require data about supplier costs and capacities, and market conditions for the company's products. Possibly, economic data about long-term prospects for the company's industry and national

economies in which the company operates its supply chain may also be required.

- *Forecast Development*
  Analytical data in the supply chain databases help to address the company's future. These data must be based on historical, transactional data. The time horizon is longest for strategic planning, but some extrapolation may be needed even for scheduling purposes (e.g., short-term forecasting of demand from large customers).
- *Parameters of Management Policies*
  The decision database must include data and structural inputs reflecting company policies and managerial judgments about risks. The decision variables and constraints that mechanize our optimization models have to be included.
- *Integration of Model Outputs with Model Inputs*
  A supply chain decision database has to include output from optimization models (like those developed in this book) as well as the data used in generating the model. Here graphical displays of model inputs and outputs are necessary, too. (This feature is also valuable for comparing and contrasting plans for multiple scenarios.)

The models developed in the earlier chapters are part of initiatives to move down the hierarchy to develop and use optimization modeling systems. Sustainable competitive advantage can only be achieved if IT innovations are combined with complementary organizational and business initiatives as well as a proper linkage to optimization models for production planning.

In the same spirit as shown in Figure 10.3 efforts have been made to group the different tasks and items into a supply chain planning matrix (see Figures 10.4 and 10.5 taken from, and with detailed descriptions, in Fleischmann and Meyr (2003) and some of the contributions in Stadtler and Kilger (2005), Stadtler (2005)). The planning tasks are ordered according to the supply chain processes procurement, production, distribution, and sales in one dimension of the matrix and according to long-term, mid-term, and short-term decisions in the other dimension. Between the entries of the matrix there is a wealth of information and material flows that go along similar lines as we have seen in the system hierarchy above.

While Figure 10.5 refers to specific tasks to be undertaken in supply chain management, Figure 10.4 provides a close linkage to respective systems and can be seen as closely related to the supply chain system hierarchy described above (see also Figure 10.3). It should be noted that some of the planning functionalities described in the matrix in fact also include scheduling aspects. Moreover, it should be noted that the systems may be completely different depending on specific industries (so that some authors propose to have a third dimension for the matrix). Computerized planning tools as they can be deduced from the supply chain planning matrix or from the hierarchy

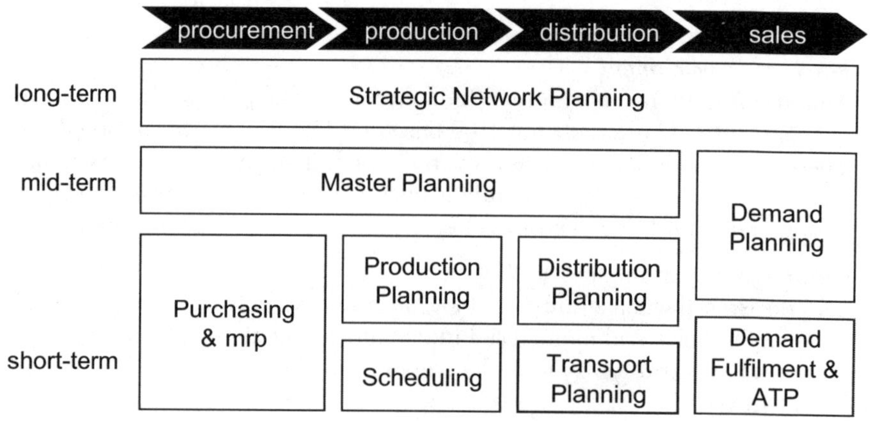

Fig. 10.4. Supply Chain Planning Matrix – System Hierarchy

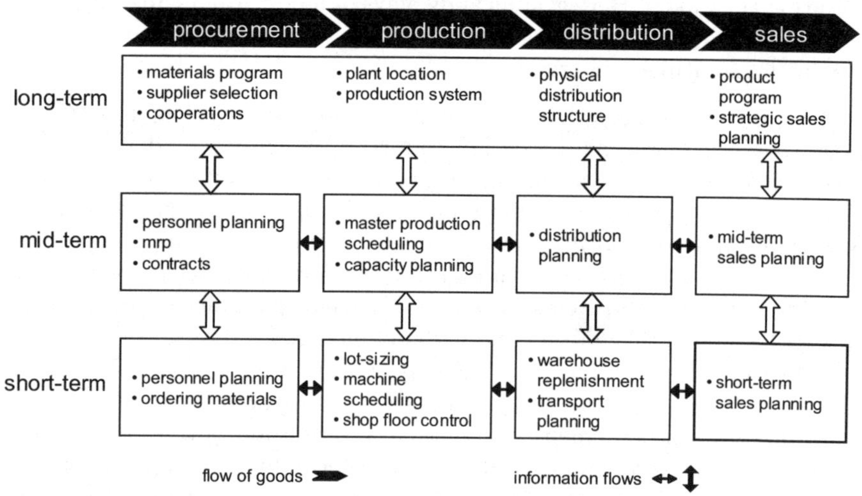

Fig. 10.5. Supply Chain Planning Matrix – Tasks

described above are often called advanced planning systems (APS). That is, APS are computerized planning tools aiming at supporting the various planning processes within supply chains.

Order management refers to management and control of customer orders from the very first customer inquiry to the finished product delivery. Once a customer order enters the systems of a company demand fulfillment is assumed to support taking care about it. Order promising refers to decisions about the acceptance of orders and setting due dates for incoming orders. Matching demand and supply asks for appropriate allocation of already accepted but yet incomplete orders with respect to yet unassigned stock as well as projected supply. Existing inventory as well as projected production that are not yet assigned to a specific customer order may be used for demand fulfillment; we say that they are "available to promise" (ATP). Beyond ATP quantities one may also consider free or unused capacities of production resources; they are "capable to promise" (CTP). For an in-depth discussion of ATP and CTP see Fleischmann and Meyr (2004).

Many companies use modern information technology to help them gain competitive advantages in the marketplace. The rapid IT advancements have provided tools to enable supply chain partners to share information with each other. Yet, questions concerning the benefits that can be gained through the sharing of information are frequently raised. Several researchers have examined the impact of information sharing on business performance; see, e.g., Lee et al. (2000), Thonemann (2002), Zhao et al. (2002). A survey on the impacts of sharing information including classification of well over 100 references is provided by Huang et al. (2003), another comprehensive treatment is given by Chen (2003).

One of the most common of information sharing issues is the so-called bullwhip effect. It describes the effect that is observed when forecasts for intermediate SKUs within a supply chain are based only on the demand experienced for those SKUs: variability in the demand pattern is magnified. That is, the further "away" we are from customer demand the more volatility is observed. A mitigating solution is to treat intermediate SKUs as dependent items by basing their forecast on the forecast for end items. This is exactly what our models prescribe for intermediate items; however, it is challenging to deploy such models across enterprise boundaries. For references on the bullwhip effect see, e.g., Lee et al. (1997), Chen et al. (2000), Simchi-Levi et al. (2002). It is also closely related to inventory control; see, e.g., Gavirneni et al. (1999) and §10.3.2. Food for thought is provided by Daganzo (2003). He describes control methods for eliminating bullwhip related instabilities without increasing supplier costs and presents approximate cost formulas.

Various games and interactive simulations have become an important part of the pedagogy of supply chain management. They are used to help explain the material and information flows in production and distribution systems. For example, the bullwhip effect is aptly demonstrated by the so-called beer game. Based on these games and the delivery of exercises showing the details and complexity of production and distribution planning, an improved understanding of supply chain issues may be demonstrated; see, e.g., the contributions in Johnson and Pyke (2000). Moreover, the field of production and operations management up to supply chain management reveals many thought provoking issues related to developing teaching material and teaching cases; see, e.g., Kanet and Barut (2003).

While software for supply chain management and enterprise resource planning is still lacking some type of planning functionality the transactional data issues have greatly advanced over the last couple of years. Related hints can be found, e.g., in Knolmayer et al. (2002), Stadtler and Kilger (2005). Finally, we mention Geunes and Pardalos (2003), who provide an annotated bibliography on the extent to which network optimization approaches have contributed to the advancement of supply chain management and financial engineering research.

10.1.4 Supply Contracts

Supply chain integration refers to the connection of at least two parties within a supply chain or network. In addition to activities associated with supply chain management, integration refers to the fact that the parties have to agree upon the way they interact with each other. Especially in the economics literature there is a long history of analysis of the contractual treatment of relationships; see, e.g., Tirole (1988), Katz (1989). Within the modern supply chain discussion the importance of the topic has not diminished but is gaining even more prominence, especially when system-wide optimization is taking place and the incentives discussion as well as concepts like transfer pricing are considered. Supply chains are, by their very nature, based on partnerships. Products as well as the data necessary for an optimization must be exchanged by the partners. What a supply contract may add to this is the explicit specification of a relationship by articulating efficiency measures or metrics that are direct input to our models such as, e.g., lead times or capacity bounds.

A more normative examination of contracts is provided in the operations research literature. For example, Bassok and Anupindi (1997) examine optimal ordering policies for a buyer where there is a pre-specified annual minimum order quantity. Developing contracts also comes along with negotiation processes between supply chain partners. Consider a supplier and its retailer. Due to the supplier's commitments with other customers the negotiation could be, e.g., about the maximum order quantity the retailer can order at a certain price; see, e.g., Homburg and Schneeweiss (2000).

Since various aspects come to mind when dealing with contracts we provide a classification scheme for supply chain contracts following Tsay et al. (1999), Voß and Schneidereit (2002) as well as the literature given there. While the classification could be developed along the lines of timing, pricing, quantity, and quality we follow the idea of having contract clauses in the foreground.

- *Specification of Decision Rights*
  Decision rights have to be defined in order to make a contract executable. Using different types of data and information, they constitute a determination of who is allowed to make decisions and within which range of action.

Control mechanisms may be centralized or decentralized as well as global versus local.

- *Information*
  Supply partners have to mutually agree about which data and information have to be exchanged at what time and through which channels.
- *Pricing (including Incentives)*
  Pricing refers to the specification of financial terms of the supply partners. Commitments have to be made regarding most aspects of the contract such as production costs or retail prices but also the dynamic aspects of cost functions (e.g., allowing for discounts in certain cases, having modified pricing on different lead times based on alternate routings). We include also the implementation of mechanisms to divide profits based on cooperation as well as the arrangement for incentives.
- *Bounds on Purchase Commitments*
  Upper and lower quantity bounds for the purchase of goods or products have to be specified. This also includes terms of flexibility, e.g., regarding early or late delivery as well as deviations from previously planned quantity estimates.
- *Allocation*
  Defines mechanisms for allocating goods in cases of limited availability.
- *Timing (including Lead Times)*
  The time of delivery of items or parts has to be specified. The lead times have to be determined and controlled. When linked with transportation clauses, this includes possible definition of push or pull mechanisms for the ordering processes.
- *Transport*
  The contract can include clauses on how the delivery is performed (e.g., by using third party logistics provider) including the definition of penalties for modifications or late arrivals as well as items damaged during transport. This is related to the implementation of rules for having various transport possibilities enabling, e.g., expedited delivery in cases of necessary adjustments in the lead times as described in §6.1.
- *Quality*
  Thresholds for the quality of the parts or items as well as allowable modifications like possibilities for upgrades or price reductions in case of unavailability of desired items are articulated in the contract.
- *Buybacks or Return Policies*
  Responsibilities for unsold inventory or products with a different quality than agreed upon have to be determined.

All these factors may become part of supply contracts. And yet, soft factors that are beyond our focus on optimization are important issues not to be neglected. This includes language skills along multinational supply chains, cultural differences as well as legal matters. From the optimization perspective we might add complications such as supply chain structures that may be characterized as one-to-one, one-to-many, many-to-one, or many-to-many. For some references see the bibliography in Cachon (2003).

10.2 mrp, MRP II and Beyond

Starting with basic concepts and then extending, we have provided an incremental approach to building optimization models for production planning in supply chains. Somehow this goes along the evolution of software available in this field. While beginning with optimization "at your fingertips," i.e., doing everything by hand, we now see a tremendous success of software vendors slowly entering the field of real planning functionality.

10.2.1 The Early Steps

Orlicky is widely credited with having "invented" mrp, or at least with popularizing it. The second edition of his seminal work on mrp is Orlicky (1975). This book explains the methods and associated record keeping needed for mrp. Bear in mind that mrp was a tremendous improvement over older management systems that were better suited to a make-to-stock environment. Shorter product life cycles and make-to-order environments require a planning system that anticipates the need for varying mixes of components.

To make better use of mrp, deeper understanding of the relationships between inventory and lead time was needed. Early work by Wight (such as Wight (1974)) helped make mrp successful. In fact, Wight is often credited with inventing MRP II as a way to make mrp logic work correctly.

The actual practice of MRP II was, and is, invented and reinvented by the software firms, consultants, planners and schedulers who make it work. A number of books and a large number of articles provide practical tips for implementing and using MRP II such as Wallace (1990) and Luscombe (1993). In such works, MRP II is referred to as a *closed loop* production planning system because the capacity check is followed by adjustments to the data followed by another execution of mrp and so forth.

A classic text by Vollmann et al. (1988) places mrp, MRP II and the associated planning tools in a broader perspective of planning and scheduling tools popularized in the 1970's and 80's. At the time of their second edition in 1988, mathematical programming approaches to production planning were considered to be an advanced concept. Simultaneous consideration of an objective function along with the materials requirements constraint and the capacity constraint was treated as part of the human aided processing of MRP II. The book presents a number of sophisticated and practical methods for planning as well as scheduling and forecasting.

At about the same time, some of the shortcomings of the overall philosophy of MRP II were beginning to be discussed (see, e.g., Kanet (1988), Spearman et al. (1990)). This process is on-going (see, e.g., Drexl et al. (1994), Spearman and Hopp (1998)) in the academic literature. The task of bridging the gap between verbally describing the philosophy of mrp and MRP II and deriving mathematical models by means of simple objective functions and constraints has been undertaken by Voß and Woodruff (2000).

Some lines of research address fundamental issues that determine the difficulty of production planning and the successful execution of a plan. For example, there is a large literature concerning SMED, where the seminal work is Shingo (1985). Another vein of research follows Goldratt and Fox (1986) and explores issues related to identifying and managing bottlenecks in production facilities and business in general.

10.2.2 Supply Chain Planning

Based on the supply chain definition on page 187, even if we speak about a chain, we really have or could have a network in mind. Among the more strategic questions in this respect are those related to network design and location. An early reference is Cohen and Lee (1988). More recently, Santoso et al. (2005) consider large-scale supply chain network design problems under uncertainty and discuss a framework for identifying and testing a variety of candidate design solutions. In §10.4 we consider a location-allocation problem. For a review of integrated strategic and tactical models and design issues see Goetschalckx et al. (2002).

Extending simple requirements planning for mrp and beyond is an important topic in the literature. For instance, Graves et al. (1998) study models for requirements planning in multi-stage production inventory systems. They develop a mathematical model to capture many of the planning issues arising in common industrial settings. The idea is to have a planning model for a single stage system as a building block and to extend appropriately. Incorporating locational decisions into a supply chain planning model encounters cross-facility capacity management. For some problems a multi-commodity flow network formulation may be used as a modeling concept; see, e.g., Wu and Golbasi (2004).

Meanwhile, a literature surrounding new technologies from ERP and supply chain management software vendors is beginning to appear (see, e.g., Gumaer (1996) as well as the references in §10.1). One of the evolving products related to supply chain planning is the Advanced Planner & Optimizer from SAP; a detailed discussion and implementation details can be found in Knolmayer et al. (2002), Dickersbach (2004). Especially in these technologies we believe that solvers based on heuristic search and constraint programming should play, and will play, a prominent role.

A final comment refers to the widespread use of ERP systems all over the world. While common understanding is that mathematics and respective models are universal, many other things like culture or language are not. A thought provoking question for software vendors refers to the markets and the possible use of ERP systems. As supply chains become global we continue to encounter boundaries that are literally beyond planning in our sense. Interesting entries into some literature, e.g., considering questions of the use of mrp, MRP II, and ERP systems in, say, China are Wang et al. (2005), Zhao et al. (2002).

10.3 Production Planning and Scheduling

The models that we have introduced are primarily oriented toward planning for production, but as we noted plans must be constructed with an eye toward the eventual creation of a corresponding schedule. There is a large body of academic literature concerning planning, scheduling and closely related topics. In the subsections that follow we provide some connections to this literature that can be used by the interested reader to gain access to these lines of research. There are a number of outstanding texts devoted to production planning and scheduling such as Johnson and Montgomery (1974), Hopp and Spearman (2000), Nahmias (2004) and, in German, Domschke et al. (1997). The citations in these books also provide a good entry point for further study of the academic literature.

10.3.1 Lot Sizing Models

A Classification Scheme. Lot sizing problems can be characterized by a variety of aspects and classification criteria. The most important distinction refers to deterministic versus stochastic models. While in deterministic models all data are known in advance, in stochastic models data are based on distributions or a measure of uncertainty.

Static models assume that parameter values do not change over the planning horizon (e.g., a continuous demand at the same rate in every period) while dynamic models allow for variation. The planning horizon can be assumed to be finite or infinite. Some of the most important data within lot sizing models are cost data. They may refer to various sorts of cost, such as, e.g., holding costs, setup costs, or production costs.

For the number of products we distinguish between models that consider exactly one and those which take multiple products into account. The latter may imply the difficulty of having to provide plans for these products on several scarce resources. In multi-stage models other than in single-stage models one considers a given product structure based on given interdependencies between the products as we have used it when defining data $R(i,j)$ based on a bill of materials.

An important distinguishing characteristic of lot sizing formulations is capacity modeling. Capacitated models recognize that some resources are given in a limited number or amount so that planning and scheduling systems need to avoid overutilizing these resources. In situations were there is not enough capacity, one might consider producing or ordering goods after they are actually needed. In this respect one distinguishes between backorder, when this indeed is possible (while paying, e.g., some sort of delay costs), and lost sales (i.e., where the customer refuses to accept any produced items after a given due date).

While in reality we are facing some finite production times, academics often assume that they are able to produce infinitely fast. Depending on the objectives this simplification may make sense.

To exemplify concepts we discuss some modeling aspects regarding a specific lot sizing problem in more detail.

The Capacitated Lot Sizing Problem. The capacitated lot sizing problem (CLSP) in its original form is a simple to state and yet difficult to solve dynamic lot sizing problem that is very similar to our "better" MRP II model (see model **SCPc** in §5.5). To start with a simple version of the models that appear in the research literature, we assume that we have a set of P SKUs that are to be produced within T time buckets.

As in §3.4, we have decision variables, $x_{i,t}$, which specify the quantity of SKU i to be produced in period t. That is, these variables indicate the lot sizes which may change over time. Whenever production of an SKU i takes place in any period we have to pay a setup cost which will be denoted by $C(i)$ using the same cost data as in §5.1.1. In order to enforce the payment of the setup cost, we use an indicator variable, $\delta_{i,t}$, that will be one if any of SKU i will be produced in period t.

To simplify further we assume that there are no lead times, and there is no bill of materials, i.e., all $R(i,j)$ will be 0 and, therefore, omitted from the model. As an important part of the objective function we have to pay a holding cost $H(i)$ for every unit of SKU i that is kept in stock. Inventory of SKU i at period t will be denoted by $I_{i,t}(x,\delta)$ with $I_{i,0}(x,\delta) = I(i,0)$ indicating the beginning inventory of SKU i and $D(i,t)$ the external demand for SKU i in period t. Then the demand and materials requirement constraints for all $t = 1, ..., T$ and $i = 1, ..., P$ read as follows:

$$\sum_{\tau=1}^{t} x_{i,\tau} + I_{i,t-1}(x,\delta) - \sum_{\tau=1}^{t} D(i,\tau) - I_{i,t}(x,\delta) \geq 0$$

The meaning of these constraints refers to the fact that in each period we need to have enough inventory from the previous period and enough made in that period to fulfill the demand. Note that one of the assumptions of the CLSP is that lead times are not explicitly considered, i.e., any production $x_{i,t}$ is available within period t. Whatever remains goes over to the next period as inventory. Equivalently, we could have used the following set of constraints:

$$x_{i,t} + I_{i,t-1}(x,\delta) - D(i,t) - I_{i,t}(x,\delta) \geq 0 \qquad i = 1,\ldots,P,\ t = 1,\ldots,T$$

The modeling constraint for the production indicators is: $\delta_{i,t} \geq \frac{x_{i,t}}{M}$. As an art of modeling we ask ourselves how small M could be to do what it is supposed to do? One guess refers to the amount yet to be produced, i.e.,

$$M = \sum_{\tau=t}^{T} D(i,\tau).$$

Furthermore, we have the integer constraint for the production indicator $\delta_{i,t} \in \{0,1\}$ and the non-negativity of the production $x_{i,t} \geq 0$.

As the CLSP is a capacitated problem, the last thing to take care of is the available capacity. The capacity constraints typically used in the literature can be rewritten to match those used in our models. Let $U(i,t)$ denote the fraction of available time needed to make one unit of SKU i. Then we have the "capacity constraint" $\sum_{i=1}^{P} U(i,t)x_{i,t} \leq 1$ for all time buckets t. However, authors in the CLSP literature often use slightly different notation and typically refer to time as the scarce resource. In other words, for them $U(i,t)$ represents the fraction of the time bucket consumed by one unit of SKU i.

To summarize, the CLSP is given in Figure 10.6

Minimize:

$$\sum_{t=1}^{T}\sum_{i=1}^{P} [H(i)I_{i,t}(x,\delta) + C(i)\delta_{i,t}]$$

subject to:

$$\begin{array}{lll}
x_{i,t} + I_{i,t-1}(x,\delta) - D(i,t) - I_{i,t}(x,\delta) & \geq 0 & i = 1,\ldots,P,\ t = 1,\ldots,T \\
\delta_{i,t} - \frac{x_{i,t}}{M} & \geq 0 & i = 1,\ldots,P,\ t = 1,\ldots,T \\
\sum_{i=1}^{P} U(i,t)x_{i,t} & \leq 1 & t = 1,\ldots,T \\
\delta_{i,t} & \in \{0,1\} & i = 1,\ldots,P,\ t = 1,\ldots,T \\
x_{i,t} & \geq 0 & i = 1,\ldots,P,\ t = 1,\ldots,T
\end{array}$$

Fig. 10.6. CLSP Model

A Modification. The CLSP generally considers P products that are not linked by means of a BOM. As a matter of modeling variety we should note that there is a possibility to define sets regarding the BOM, i.e., *Pred*(i) as the set of predecessors of SKU i and $Succ(i)$ as the set of successors of i. Then

$$\sum_{j=1}^{P} R(i,j)x_{j,\tau} \qquad \text{can be replaced by} \qquad \sum_{j \in Succ(i)} R(i,j)x_{j,\tau}.$$

We will now consider a special situation where there is exactly one end-item and the product structure is strictly convergent, which means that each of the other SKUs is a component in only one subsequent SKU. Keeping the numbering of the SKUs in a low-level-coding we assume a BOM where every SKU but the first has exactly one successor (like in the simple example in Figure 3.1 in §3.1). In such a case we name this successor of i by $succ_i$. The product structure may be viewed as convergent because there is only a single end item, which is SKU 1. As data in this restricted case we have $R(i, succ_i)$ indicating the quantity of SKU i needed to make one $succ_i$. Consider the problem in Figure 10.7.

Minimize:

$$\sum_{t=1}^{T}\sum_{i=1}^{P} [H(i)I_{i,t}(x,\delta) + C(i)\delta_{i,t}]$$

subject to:

$$
\begin{array}{lll}
x_{1,t} + I_{1,t-1}(x,\delta) - D(1,t) - I_{1,t}(x,\delta) \geq 0 & & t = 1,\ldots,T \\
x_{i,t} + I_{i,t-1}(x,\delta) - D(i,t) - R(i,succ_i)x_{succ_i,t} - I_{i,t}(x,\delta) & \geq 0 & \\
 & & i = 2,\ldots,P,\ t = 1,\ldots,T \\
\delta_{i,t} - \frac{x_{i,t}}{M} & \geq 0 & i = 1,\ldots,P,\ t = 1,\ldots,T \\
\delta_{i,t} & \in \{0,1\} & i = 1,\ldots,P,\ t = 1,\ldots,T \\
x_{i,t} & \geq 0 & i = 1,\ldots,P,\ t = 1,\ldots,T
\end{array}
$$

Fig. 10.7. A Lot Sizing Model with Convergent Product Structure

As an advanced exercise one could verify that this model is an example of the **SCPc** model on page 57 under very special conditions. This problem is interesting from a modeling standpoint as pointed out by Afentakis et al. (1984) and Domschke et al. (1997). For instance, we may use a general variable redefinition approach following Martin (1987). While staying with the same data as above as well as the variables $\delta_{i,t}$ as production indicator for production of SKU i in period t and $I_{i,t}(x,\delta)$ to denote the inventory of SKU i at period t some additional binary variables will be defined. We call them availability variables and they are denoted by $z_{i,\tau,t}$ indicating, if they take the value 1, that the demand for SKU i in period t is produced in some period from the first period up to period τ. With these binary variables we have a nice way of re-formulating our problem.

Let us discuss the availability variables in more detail. Whenever the demand $D(i,t)$ is produced in some period $\tau^* \in \{1,\ldots,t\}$ then $z_{i,\tau,t} = 0$ for $\tau \in \{1,\ldots,\tau^*-1\}$ and $z_{i,\tau,t} = 1$ for $\tau \in \{\tau^*,\ldots,t\}$. That is, $z_{i,\tau,t} = 1$ indicates the (systemwide) availability of that demand.

We have to ensure that the demand for any period t is available not later than in that period, i.e.:

$$z_{i,t,t} = 1 \qquad i = 1,\ldots,P, \quad t = 1,\ldots,T$$

Furthermore, one has to guarantee that the inventory for an SKU, once available, does not get lost:

$$z_{i,\tau+1,t} - z_{i,\tau,t} \geq 0 \qquad i = 1,\ldots,P, \quad \tau = 1,\ldots,t-1, \quad t = 1,\ldots,T$$

Once the left hand side of this constraint is one, this indicates a change in the availability and hence the production indicator needs to be forced to 1:

$$z_{i,\tau+1,t} - z_{i,\tau,t} \leq \delta_{i,\tau+1} \qquad i = 1,\ldots,P, \quad \tau = 1,\ldots,t-1, \quad t = 1,\ldots,T$$

Finally, we have to consider the BOM which is a convergent product structure for this problem with a single end item, which is SKU 1:

$$z_{i,\tau,t} - z_{succ_i,\tau,t} \geq 0 \qquad i = 1,\ldots,P, \quad \tau = 1,\ldots,t-1, \quad t = 1,\ldots,T$$

The objective function has to consider all relevant costs over the planning horizon which are setup costs and holding costs.

$$\text{minimize:} \sum_{t=1}^{T}\sum_{i=1}^{P} C(i)\delta_{i,t} + \sum_{t=1}^{T}\sum_{\tau=1}^{t-1} H(1)D(1,t)z_{1,\tau,t}$$
$$+\sum_{i=2}^{P}\sum_{t=1}^{T}\sum_{\tau=1}^{t-1} H(i)D(i,t)\left[z_{i,\tau,t} - z_{succ_i,\tau,t}\right]$$

The second term of the objective function considers holding costs for finished amounts of the end item. Whenever an SKU is available but has not yet been used for the production of its successor we have to account for the holding costs as is done in the third term of the objective function. Without loss of generality we can assume that there is a demand for all SKUs at the first time bucket.

With this we can summarize the model in Figure 10.8. The solution to this problem directly implies the values for $x_{i,t}$. The reformulated version of the problem is computationally attractive because the problem has only binary variables and binary constraint coefficients. The use of big M has been eliminated.

Minimize:

$$\sum_{t=1}^{T}\sum_{i=1}^{P} C(i)\delta_{i,t} + \sum_{t=1}^{T}\sum_{\tau=1}^{t-1} H(1)D(1,t)z_{1,\tau,t}$$
$$+\sum_{i=2}^{P}\sum_{t=1}^{T}\sum_{\tau=1}^{t-1} H(i)D(i,t)\left[z_{i,\tau,t} - z_{succ_i,\tau,t}\right]$$

subject to:

$$\begin{array}{lll}
z_{i,t,t} & = 1 & i = 1,\ldots,P,\ t = 1,\ldots,T \\
z_{i,\tau+1,t} - z_{i,\tau,t} & \geq 0 & i = 1,\ldots,P,\ \tau = 1,\ldots,t-1,\ t = 1,\ldots,T \\
z_{i,\tau+1,t} - z_{i,\tau,t} & \leq \delta_{i,\tau+1} & i = 1,\ldots,P,\ \tau = 1,\ldots,t-1,\ t = 1,\ldots,T \\
z_{i,\tau,t} - z_{succ_i,\tau,t} & \geq 0 & i = 1,\ldots,P,\ \tau = 1,\ldots,t-1,\ t = 1,\ldots,T \\
z_{i,\tau,t} & \in \{0,1\} & i = 1,\ldots,P,\ \tau = 1,\ldots,t-1,\ t = 1,\ldots,T \\
\delta_{i,t} & \in \{0,1\} & i = 1,\ldots,P,\ t = 1,\ldots,T
\end{array}$$

Fig. 10.8. A Reformulation of the Lot Sizing Model from Figure 10.7

Further Dynamic Lot Sizing Models. A paper by Billington et al. (1983) is the first that we know of to propose a model like **SCPc**. They also provide some guidance for problem reduction based on the bottleneck (see §6.6). The problem is viewed directly from the MRP II perspective by Adenso-Díaz and Laguna (1996) where the model includes the possibility of overtime. The objective is to minimize the use of overtime to meet the demand requirements. Other works, such as Tardif and Spearman (1997) take a more algorithmic approach. This work focuses on capacity constrained mrp systems and provides a methodology for finding capacity feasible production plans.

Many of the planning and scheduling models proposed in the research literature are labeled as lot sizing. We have argued against lot sizing, but really we oppose only *static* lot sizes. The literature on dynamic lot sizing is primarily about scheduling, and this can be quite sensible.

An excellent example of this literature is a paper by Trigeiro et al. (1989), which addresses a variant of the CLSP. Their assumptions are consistent with the CLSP in that they ignore sequencing. They assume that production of all parts from a family in a period is done in one batch that requires a setup. The objective function considers holding costs (due to early completion), temporal variations in production costs, and setup costs. The model is very similar to the **SCPc** model given here except that they provide the modeling details necessary to allow production of an SKU to span two time buckets. They also provide details of a special purpose solution technique based on Lagrangian relaxation that proves to be very effective for this problem.

A great deal of effort has been put into various modeling and solution approaches for different modifications and generalizations of the CLSP; see, e.g., Ertogral and Wu (2000), Suerie and Stadtler (2003), Stadtler (2003). The CLSP may be extended in a variety of ways. Let us consider the case where time buckets are so small that in each ("micro") time bucket only one SKU can be produced. In the literature this problem is referred to as *Discrete Lot Sizing and Scheduling Problem.* Accordingly, the proportional lot sizing and scheduling problem allows for the production of at most two SKUs in one time bucket. For a survey on these and related problems see Drexl and Kimms (1997).

Besides the problems already mentioned, the lot sizing literature is quite large; see Domschke et al. (1997) for a comprehensive survey as well as, e.g., Clark and Armentano (1995), Katok et al. (1998), Tempelmeier and Derstroff (1996), Stadtler (2000). Each of these models makes different assumptions that result in a different model. All of them are intended to determine production quantities. The lot sizing models typically assume a fixed lead time of one period and do not consider alternative routings. These models are similar to the basic model **SCPc** in that they are useful for production scheduling within a factory, but not appropriate for assigning production to factories within a supply chain. Alternatively, our work can also be seen as extending the category denoted as "multiple-stage production planning with limited resources" by Simpson and Erenguc (1996). In the lot sizing literature, if the work considers multiple levels in the bill of materials, the data $R(i,j)$ are often called *production coefficients.*

Naturally, most lot sizing problems cannot be solved without taking into account capacity constraints as we have seen, e.g., for the CLSP. The capacity constraints may refer to a single machine or to non-identical parallel production lines (heterogeneous machines), just to mention some possible complications. Recent interest in solving such problems also considers the application of meta-heuristics; see, e.g., Meyr (2002). Note that often the sequencing problems are treated separately from the lot sizing problems; see §10.3.3. Depending on specific industries this makes sense while for others a simultaneous lot sizing and sequencing seems more appropriate which is in line with our discussion regarding the supply chain planning matrix on page 195.

A prototype modeling and optimization system for lot sizing problems based on the branch and bound principle is provided by Belvaux and Wolsey (2000). The user needs to formulate the problem as a MIP using Xpress-MP (based on a modeling language, see §7.5) taking into account a reserved set of key words for specific lot sizing objects.

10.3.2 Planning and Inventory Control

Planning Horizon. Often data evolve over time. Research on horizon issues focuses on quantifying the diminishing effect of future data on initial decisions. To formalize the different horizon concepts, we say that a problem has a finite (planning) horizon if a finite number of, say T, time buckets is considered for planning. Assume given numbers $1 \leq t_d \leq t_f < T$. If optimal decisions up to period t_d are independent of the data beyond t_f up to T then t_d is called a decision or planning horizon and t_f is called forecast horizon; see Bes and Sethi (1988). As issues related to planning horizons and forecasting are becoming more and more important in planning and supply chain management, a comprehensive collection of references in this area, as it is provided in Chand et al. (2002), is helpful.

Multiple Routings and Subcontractors. The presence of alternative routings and subcontractors is an important feature of supply chain planning, which has not received enough attention. Chandra and Tombak (1992) look at ways to evaluate the flexibility that is provided by alternate routings. This work is useful during the design of supply chains. A paper by Kamien and Li (1990) examines subcontracting at the aggregate level and discusses economic effects and the structure of the subcontracting relationship. They show that, under certain conditions, subcontracting reduces the variability in production and inventory. This paper, like van Mieghem (1999), provides insight into subcontracting policies but does not prescribe methods for production planning in the presence of multiple routing opportunities.

A paper more directly related to our model for multiple routings is one by Logendran and Ramakrishna (1997) who create a mathematical programming model for the problem of scheduling work in manufacturing cells that have duplicate machines available for bottleneck operations and/or subcontractors who can perform the bottleneck operations. They also give details necessary to use a general-purpose heuristic solution method described in §8.4.6.

Solution methodologies for single products with subcontracting are provided by Atamtürk and Hochbaum (2001). This work also considers the interaction between the operational decision to subcontract and tactical capacity decisions. The paper provides algorithms and insights for both aspects of the problem.

Inventory Control. The models that we have proposed in the preceding chapters are appropriate for plans based on the best information available at the time of the planning process as opposed to *policies* that are parameters for making decisions. For example, a class of inventory control policies are of the basic form Q,R where Q gives the quantity to order or to produce whenever the inventory level gets down to R. An interesting example from this literature is Sobel and Zhang (2001), where ordering policies are considered for a single product when some of the demand is best modeled as being deterministic and some is best modeled as stochastic. Although we might use the word "planning" when describing the process of setting policies, it is clearly not the same activity that we have been concerned with. However, the problem statements are similar.

A problem tangentially related to the one studied in §6.1 is the problem of setting inventory policies when there exist two supply modes with differing lead times. Whittemore and Saunders (1977) look at the problem of determining the appropriate reorder policies when there are two delivery options: one fast and expensive and the other slower and less expensive. They use a stochastic dynamic programming formulation to balance the cost of backlogging and order costs with the cost of holding inventory. Moinzadeh and Nahmias (1988) produce an extension to Q,R policies for a similar model in the continuous case, except that their model includes a cost per stockout incident rather than per unit time.

There is also a considerable literature on setting inventory control policies for an entire bill of materials simultaneously. The multi-echelon inventory and related literature provides methods for setting policies to control inventory levels for a complete production/distribution system under a variety of conditions; see, e.g., Chen (1998), Hwang and Singh (1998), Minner (2000), Roundy (1986). In the simplest case, inventory policies set reorder points that imply minimum planning levels for inventory (i.e., *safety stock*) as we mentioned in §6.3. Formulas for setting safety stock levels are contained in most operations management texts; see, e.g., Martinich (1997). Sethi et al. (2005) have recently published a book that has planning and control models for inventory and supply chain planning with uncertain demand.

One of the newer ideas related to inventory control is that of vendor managed inventory (VMI). In fact this is about partnering. Assuming a supplier and a retailer in a supply chain then VMI is about the suppler taking control over the inventory policies of the retailer as well as the decisions influencing production, distribution and shipment. An interesting study by Disney and Towill (2003) investigates the impact of VMI on the bullwhip effect. The analysis shows that with VMI implementation two sources of the bullwhip effect may be completely eliminated.

Deterioration and Perishability. Deterioration may be regarded as the process of decay, damage or spoilage of products such that they cannot be used for their original purpose anymore, i.e., they gradually undergo a change in storage and lose their utility at least partially. This is in contrast with perishable items that, at some point in time, lose all of their value. Deterioration (and perishability) need to concern supply chain managers when thinking about inventory control as well as when undertaking production planning with products eventually waiting in front of a resource in order to be processed. While classical inventory models assume that inventory can be stored indefinitely in order to meet future demands, this is not realistic for products or items subject to deterioration or perishability. Early literature on deterioration and perishability is chronicled in a survey by Nahmias (1982). A more recent collection of references is compiled by Goyal and Giri (2001).

A few additional examples of this recently expanding literature are, e.g., Benkherouf et al. (2003), Balkhi and Benkherouf (2004). Quality alteration of products can be affected by random changes in the ambient environment resulting in possible deterioration in some periods while there is no deterioration in others. Aggoun and Benkherouf (2002) consider inventory control approaches in such an environment where, additionally, prices of the items of different quality are allowed to change from period to period in a random fashion.

Hsu (2000) represents an economic lot size model for perishable inventory where stock deteriorating rates depend on the stock age as well as on their production periods. The latter seems realistic as deteriorating items may decay with variable speed at different points in time. Dye and Chang (2003) discuss an economic order quantity system that includes time varying demand and deteriorating items with conditions of permissible delay in payments. Sana et al. (2004) investigate a production-inventory model for a deteriorating item over a finite planning horizon with a linear time varying demand, finite production rate and shortages. Inderfurth et al. (2005) provide analytical insights into optimal lot sizing in a hybrid production/rework environment when both switching from production to rework and vice versa is associated with perceptible cost and time.

10.3.3 Machine Scheduling

Problems in machine scheduling are closely related to lot sizing problems, but consider a higher level of detail than is used in a planning model. The solutions to the planning models developed in the first part of this book are often used as inputs to scheduling problems. Consequently, it is sometimes necessary to solve the scheduling problems as part of the planning process in order to verify that the induced problems have a solution. A simple example of this is given in §8.4.3 where the capacity constraint is replaced with a scheduling problem. A wide variety of scheduling problems have been considered in the research literature.

Textbooks on the topic of scheduling include Baker (1974), Blazewicz et al. (2001) and Brucker (2004). Related material with a focus on manufacturing has been collected in Pinedo (2005). A survey of research concerning MIP modeling of changeovers in production planning and scheduling is provided by Wolsey (1997). A survey of research on sequencing with earliness and tardiness penalties is provided by Baker and Scudder (1990).

An important topic in machine scheduling is the creation of production sequences when there are significant setups, but only a single resource to schedule. As introduced in §8.4.2 we speak of jobs which resume a collection of one or more of the same SKU to be produced. If the jobs to be sequenced are given (perhaps by the planning process) and the time to change from one job to another depends on both jobs, then the problem of finding the fastest sequence can be modeled as the *traveling salesman problem* (TSP). The TSP is a classic problem where one is given a list of cities and the distances between them and asked to find the shortest route that visits all cities. Replacing the cities by jobs and the distances by production and changeover times, we see that the model applies to sequencing problems as well.

There are a number of other models that are more general than the single machine scheduling problem just introduced. A problem with multiple resources in series where all jobs make use of all resources one after the other is called a *flow shop problem.* To make it more practical, all sorts of modifications are treated in the literature, such as setup times or no-wait constraints. For a survey on the literature in the first case see Cheng et al. (2000).

The second case requires that an SKU or a job once finished on one resource has to be processed immediately after that on the next resource without any interruption until it has left the last resource. For those so-called continuous flow shop scheduling problems the processing of each job has to be continuous, i.e., there must not be any waiting times between the processing of any consecutive tasks regarding this job. To allow processing of a job without interruption on all resources, the order in which the jobs are processed on a resource is the same for all of them (assuming non-zero processing times). If the objective is to finish all jobs as fast as possible this modification reduces again to the TSP. If we strive for minimizing the sum of the completion times of all jobs it relates to a generalization of the TSP which is called time-dependent TSP; see Gouveia and Voß (1995), Fink and Voß (2003).

When there are multiple resources in parallel, such models no longer apply and the scheduling problems become much harder; see, e.g., Belouadah and Potts (1994), Monma and Potts (1993). Another complication is when there are multiple resources needed for each job in some order which varies from job to job. This is referred to as the job shop scheduling problem; see, e.g., Vaessens et al. (1996), Pezzella and Merelli (2000), Meloni et al. (2004). In the fully general case, these two problems are combined to create the hybrid job shop problem; see, e.g., Imaizumi et al. (1998), Gupta et al. (1997).

Both the flow shop and the job shop scheduling problem generalize to the resource constrained project scheduling problem which is concerned with scheduling a set of jobs (or activities) subject to constraints on the availability of several shared resources; see, e.g., Klein (2000). Naturally, one may incorporate temporal constraints allowing the specification of minimal and maximal time lags between two activities; see, e.g., Dorndorf et al. (2000) who consider the minimization of the maximum of the completion times of all activities. Using the resource constrained project scheduling problem together with a basic mrp model allows for a reasonable augmentation leading to options for incorporating capacity constraints and variable lead times as has been investigated, e.g., by Rom et al. (2002).

10.3.4 Aggregation and Part Families

An important form of abstraction in the planning process is the consideration of aggregated parts and SKUs as developed in §6.6. This concept has applications outside the planning process as well. Given its importance, a variety of research has been conducted concerning this topic.

Some of the earliest work on systematic formation of part families was done by Soviet engineers in the later 1950's to support application of cellular manufacturing. The collection of a group of machines into a cell to process a family of similar parts remains an important topic. There are a number of algorithms available for dividing SKUs into groups to support manufacture by different cells; see, e.g., Miltenburg and Zhang (1991), Venugopal (1999). Another important application of part families is computer assisted process planning (see, e.g., Koenig (1994)). This results in a hierarchy of families driven by a part classification scheme. The classifications are done so that it is possible to reuse process plans for similar parts when a new part is designed. Such classification schemes, therefore, are based on manufacturing characteristics and can be very useful for family formation for the purposes described in §6.6.

Our application of part family formation was for the problem of aggregate planning. This has also been the subject of a significant research literature. In the 1970's researchers at MIT collectively developed a planning system that they referred to as hierarchical production planning (HPP). The HPP system is designed to translate aggregate forecasts into part production requirements. The overall HPP approach is described and justified in a paper by Hax and Meal (1975).

The hierarchy is based on scheduling parts in increasing levels of disaggregation. At the lowest level, parts are scheduled. Before that, families of parts are scheduled (intra-family changes do not require a setup, but inter-family changes do). At the highest level are *types* which are part families grouped according to similarities in production and demand. The scheduling of types is referred to as aggregate planning and the decomposition into family and part schedules is referred to as disaggregation. Bitran and Hax (1977) suggested optimization of sub-problems for the various levels and a rigorous means of linking them. We will now discuss simplified versions of the optimization sub-problems to make their methodology more concrete.

Aggregate planning is a cost minimization problem with respect to temporal variations for production to meet demand forecasts. There are constraints to assure that demand is met, periods are correctly linked via inventory, and capacity is not exceeded. Disaggregation to families is done with the objective of minimizing the total setup *cost.* A survey of disaggregation procedures is contained in Bitran et al. (1981). The disaggregation to parts is done with the objective of minimizing setup cost subject to capacity feasibility, producing the quantities specified in the family disaggregation and keeping inventories within safety stock and overstock limits. Bitran et al. (1982) have proposed a two-stage version of the model which also schedules production of components.

Although this approach is demonstrably better than mrp in some circumstances (see Hax and Candea (1984)), it is not universally applicable. It does not consider due dates. Due dates can be an important factor as flow times drop and competitive pressures increase. Their work is perhaps most appropriate when production of components can and must be begun in response to forecast demand. A more severe problem is that the minimization of setup "costs" is not generally appropriate. In some cases there are actual costs associated with a setup such as materials and supplies. But more generally, the costs are due to lost capacity and in fact depend on the schedule and part mix; see, e.g., Karmarkar (1987).

A hierarchical planning system was also proposed by Spearman et al. (1989) for a specific type of control system known as CONWIP (see Spearman et al. (1990)), where CONWIP stands for "constant work in process." These systems proposed the use of a hierarchical system as a way of dividing the problem along sensible lines to improve the ability to solve the resulting problems, which was our goal in §6.6 as well. A good reference for hierarchical planning is Schneeweiss (1999). A somewhat extended view of aggregation can also be found in Leisten (1998). A hierarchical planning approach in the light of lot-sizing and scheduling allowing only one setup per period, i.e., at most two products are allowed to be produced in each period, is discussed by Rohde (2004).

10.3.5 Load Dependent Lead Times

Lead times are an important attribute of a product. Consequently, lead times are the subject of research into their causes and effects; see, e.g., Bartezzaghi et al. (1994), Ben-Daya and Raouf (1994), Hopp et al. (1990), de Kok and Fransoo (2003), Lambrecht et al. (1998), Lee et al. (1989), Ornek and Collier (1988), Vendemia et al. (1995). The management of lead times at the control level can be accomplished using control strategies such as CONWIP (see Spearman et al. (1990), Spearman and Hopp (1998)) or Kanban (see Hall (1983), Krajewski et al. (1987), Kimura and Terada (1981), Schonberger (1986)). Use of these methods reduces variations in realized flow times for physical parts due to congestion. But unless the planning systems take lead times into account, the effect will be to increase the waiting time for parts to enter production but not the overall flow time from order release to completion. Clearly, it is better for parts to suffer congestion delays before they have started production rather than after (they can be rerouted much more easily, for one thing) so use of CONWIP and Kanban have significant benefits. These benefits can be extended when coupled with a planning system that is lead time sensitive.

It is useful to consider these systems under the classification scheme of so-called *push* versus *pull* systems. Whereas in a pull system production is initiated as a reaction to present demand, in a push system production is performed in anticipation of future demand (see, e.g., Karmarkar (1987)). CONWIP and Kanban may be referred to as pull systems.

Our primary interest is in including lead time effects in planning models. Since congestion phenomena go along with bottlenecks causing load dependent lead times it is useful to start with a queuing model in order to obtain some approximations for the key parameters of the capacity constraint formulation or objective functions to be implemented in an aggregate planning model; see, e.g., Buzacott and Shantikumar (1993). As a side-remark we should mention that congestion may be related to heavy loading or heavy traffic, i.e., situations where the average fraction of time at which a server or processor is free is small, or where a machine has little spare capacity. For some mathematical background see, e.g., Kushner (2001).

In order to capture the relationship between system loading and waiting times some authors discuss planning models with *clearing functions*. The idea of clearing functions was introduced by Graves (1986) and further developed by Karmarkar (1987) and Srinivasan et al. (1988). Recently Asmundsson et al. (2002, 2003) employ a clearing function with the aim of modeling the non-linear dependency between lead times and WIP workload.

Zijm and Buitenhek (1996) develop a scheduling model with load dependent lead times that could be extended to a planning model but their most important contribution to our work is that they provide guidance on constructing functions for the waiting time given a loading. That is, they develop a manufacturing planning and control framework for a machine shop that includes workload oriented lead time estimates in order to account for the necessity to consider both lead time and capacity management in a management planning tool. For that purpose they suggest a method that determines the earliest possible completion times of arriving jobs with the restriction that the delivery performance of any other job in the system will not be adversely affected, i.e., that every job can be completed and delivered in time. The goal is to determine reliable planned lead times based on the workload that results in due dates for jobs that can be met and that can be implemented at a capacity planning level, serving there as an input for a final detailed capacity scheduling procedure that also takes into account additional resources, job batching decisions as well as machine setup characteristics. They use queuing network techniques to determine the mean and variance of lead times dependent on lot sizes, production mix and expected annual production volume. Their framework is partly based on the work of Karmarkar (1993a) and Karmarkar et al. (1985) as they employ for each network service station a queuing model with multiple part types.

Our model in §9.1.2 could be seen as making use of a piecewise linear *clearing function* for the tradeoff between loading and waiting time as envisioned by Karmarkar (1989b) with extensions to include multiple routings or subcontractors. Figure 10.9 depicts some possible clearing functions where the constant level clearing function corresponds to an upper bound for capacity

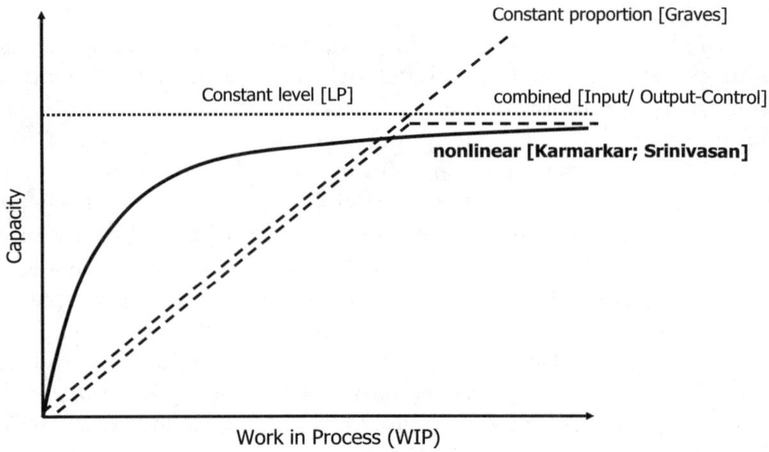

Fig. 10.9. Clearing Functions

as mainly employed by linear programming models. This implies instantaneous production without lead time constraints since production takes place independently of WIP in the production system. The constant proportion clearing function represents a control rule given by Graves (1986) which implies infinite capacity and hence allows for unlimited output. In contrast to the non-linear clearing function of Karmarkar (1987) and Srinivasan et al. (1988), the combined clearing function in some region underestimates and in others overestimates capacity. Moreover, the non-linear clearing function relates WIP levels to output and lead times to WIP levels which are influenced by the behavior of load dependent lead times. Additionally, the slope of the clearing function represents the inventory turn with lead times given by the inverse of the slope (see Karmarkar (1989a)). Special properties of the clearing function allow for formulating a linear programming version in order to develop a model which remains numerically tractable and, therefore, can deal with the problem of combinatorial explosion of company or supply chain size. The clearing function model reflects the characteristics and capabilities of the production system better than models using fixed planned lead times (like mrp).

Lautenschläger and Stadtler (1998) have a well developed method for incorporating load dependent lead times in a capacitated lot sizing model. They base their model on the standard CLSP assumption of a lead time of one period, but their model automatically delays the planned completion if the capacity is heavily utilized. Their work includes guidance on constructing functions for the waiting time given a loading.

Jagannathan and Juang (1998) describe a model where the lead times depend on lot sizes ordered. In this paper, the production level of each SKU is considered separately for the purpose of determining lead times rather than modeling shared resources and routings. They report on a solution method specifically designed for their formulation. Enns (2001) describes simulation experiments that link static lot sizes, lead times and the performance of an mrp system.

Voß and Woodruff (2004) use a piecewise linear clearing function as suggested by Karmarkar (1989a) in order to model the dependency of lead (or waiting) times in their tactical planning model including multiple routing and subcontractors and highlight the fact that lead times are dependent on the decisions considering the utilization of production resources. Consequently, in order to create realistic, feasible and robust production plans, it is imperative to integrate the effects of load dependent lead times into the tactical planning models. Research issues still remain on how high the utilization of resources could be before being considered a bottleneck and on finding more sophisticated approximations for the clearing function.

We will now give an overview of approaches to formulate clearing functions in order to capture the non-linear relationship between lead times and workload of the production system. A more comprehensive survey is compiled in Pahl et al. (2005).

Graves (1986) studied the extent to which the job flow time (or WIP inventory) depends on the utilization of each resource of a job shop or production stage. The production system is modeled as a network of queues with multiple routing and planned lead times as the decision variable. The clearing function serves as a release control rule at each resource which determines the amount of work performed during a time period which is a fixed portion of the queue of work remaining at the start of the period. As this formulation implies infinite capacity other functional formulations are suggested by, e.g., Karmarkar (1989a,b, 1993b), Srinivasan et al. (1988), Zijm and Buitenhek (1996), Missbauer (2002), Lautenschläger and Stadtler (1998), Asmundsson et al. (2002, 2003), Hwang and Uzsoy (2004), Caramanis and Anli (1999) and Mendoza (2003).

As opposed to Graves (1986), Karmarkar (1989a) and Srinivasan et al. (1988) model the non-linear relationship deriving a clearing function of the following form:

$$\text{Capacity} = \alpha(WIP) \cdot WIP$$

Here, the clearing factor α specifies the fraction of the actual WIP which can be completed, i.e., "cleared" by a resource in a given time period. Missbauer (2002) refers to this factor as the "utilization factor." In order to give an idea how the clearing function "works" in an aggregate production planning model we refer to the model of Asmundsson et al. (2002) as a reference.

The mathematical programming approach of Asmundsson et al. (2002) models the non-linear dependency between lead times and WIP (workload) by employing a clearing function, too. Special properties of the clearing function allow for formulating a linear programming version in order to develop a model which remains numerically tractable. In accordance with the procedure of Karmarkar (1989a), Asmundsson et al. (2002) define the performance of a resource (work center) as dependent on the workload. For that reason they use a queuing model including variation coefficients for the service time and the arrival time. Moreover, the utilization of a resource is formulated as a function of the WIP. Also batching and lot sizing have an effect on lead times especially when small batches give rise to frequent setup changes leading to time losses for production, lower throughput and eventual starvation of resources on further production stages.

In order to develop the clearing function, there are two methods available in the literature to date, where the first is the analytical derivation from queuing network models and the second an empirical approximation using a functional form which can be fitted to empirical data. Because of the large amount of details in practical systems the complete identification of the clearing function will not be possible, so we have to work with approximations. Asmundsson et al. (2002) integrate the estimated clearing function in a mathematical programming model where the framework is based on the production model of Hackman and Leachman (1989) with an objective function that minimizes the overall costs. It is assumed that backorders do not occur and that all demand must be met on time. In contrast to Ettl et al. (2000) the non-linear dynamic is incorporated in the clearing function in the constraints and thus not included in the objective function. For more detail see Asmundsson et al. (2003).

Modifications of batch sizes can be a good instrument to control the workload in the system since workload, WIP, safety stocks and lead times (or flow times) are dependent on the choice of the batch size; see, e.g., Zipkin (1986), Karmarkar (1989a), or Karmarkar et al. (1985).

Karmarkar (1989a) develops a capacity and release planning model which explicitly takes into account WIP costs and lead time consequences caused by the production system workload. For that purpose it is based on order releases and batching and applies the traditional capacity planning methodology that combines release planning, master scheduling issues and seasonal planning. Additionally, it aims at surmounting the shortcomings of aggregate production planning models like those of Graves (1986), Kekre and Kekre (1985) and the limitations of input/output - control models. Like Srinivasan et al. (1988), Karmarkar (1989a) uses the non-linear "clearing function" to represent the output as a function of the average WIP in the production system. The form of the curve is also valid for synchronous deterministic flow lines with batched flows. In order to keep things simple, Karmarkar (1989a) considers a discrete period model for a single product production system to proceed to the dynamic reformulation of the model.

Hwang and Uzsoy (2004) combine the work of Karmarkar (1987) and Asmundsson et al. (2002) and add lot sizing in order to show how small or large lot sizes influence the resulting production plans. For that purpose they present a single-product dynamic lot sizing model which takes into account WIP and congestion using queuing models such as those presented by Karmarkar (1987, 1993b) and develop a clearing function which captures the dependency of the expected throughput of a single-stage production system closely related to the classical Wagner-Whitin model (see Wagner and Whitin (1958)) including setups, expected WIP levels and lot sizes which is then integrated in a dynamic lot sizing model. Results demonstrate that their proposed model provides significantly more realistic performance and, therefore, production plans than models ignoring the relationships between lead times, workload, throughput, production mix and lot sizes (setups).

Before closing this section we should mention that research on workload control has had a wealth of interest especially in the semiconductor industry. This area has developed and considered various workload control concepts investigating general dispatching and order release methods with a focus on wafer fabrication, i.e., the combination of lot release and dispatching strategies used to control the flow of lots through a semiconductor wafer fabrication facility. Uzsoy et al. (1994) provide a general survey and review of production planning and scheduling models in semiconductor manufacturing. A more recent survey with a clear focus on workload control is provided by Fowler et al. (2002). Moreover, we like to point to some references that we feel provide some innovative or thought provoking ideas in one sense or another: Hackman and Leachman (1989), Hung and Leachman (1996), Leachman (1993), Schoemig (1999).

10.4 Transportation

Product movement and transportation is an important part of supply chain management. Transportation modeling may be concerned with routing SKUs between different machines (see §10.3.3) or between a variety of locations. We have proposed extensions to our planning models for some aspects of transportation planning, see §6.4. Within supply networks we are also concerned about shipping, which in our models was assumed to be included in the lead time without careful consideration of how to control it. Most related problems are those "above" our models, i.e., building up a transportation system, and those "below," i.e., up to now the real movement had been left out on purpose. Much has been written about pure transportation problems as well as on various generalizations; see, e.g., the surveys and collections of Laporte and Gendreau (1995), Cordeau et al. (1998), Kwon et al. (1998), Hall (2003).

A family of models has been developed for optimization of the transportation of goods when production and demand quantities as well as the locations of factories, distribution centers and customers are known in advance. This is a classic optimization model as the basis for teaching and understanding the application of optimization models in addition to being a useful transportation model. One form of the transportation problem is given in Figure 10.10

where the decision variables $t_{i,j}$ are the quantity of SKUs to transport or ship from i to j and the data are the cost to transport or ship from i to j, $T(i,j)$, the demand at j, $D(j)$, and the capacity at i, $C(i)$. Often all i's are called origins and all j's destinations. The number of origins is L and the number of destinations is N.

Minimize:

$$\sum_{i=1}^{L}\sum_{j=1}^{N} T(i,j)t_{i,j}$$

subject to:

$$\sum_{i=1}^{L} t_{i,j} = D(j) \qquad j = 1,\ldots,N$$

$$\sum_{j=1}^{N} t_{i,j} \leq C(i) \qquad i = 1,\ldots,L$$

Fig. 10.10. Transportation Problem

While minimizing the cost of all shipped quantities, we have two sets of constraints that guarantee that all the demands are fulfilled and that all the capacity is not overused. This problem and many of its variants have been thoroughly studied over the years and can be solved efficiently. See, e.g., Clarke and Wright (1964), Hall and Racer (1995), Solomon (1987).

This model has been the basis for a plethora of more realistic transportation models. For example, these include the binary transportation problem or the more general design of transportation systems; see, e.g., Belenky (1998), Bhaskaran and Turnquist (1990), Fleischmann (1998), Fleischmann et al. (2001).

These transportation models find optimal quantities, but do not seek to specify operational details such as delivery routes. Shipment cost data are typically based on averages. When we consider models for optimal route planning, we once again encounter the TSP, which is a classic both within transportation planning as well as optimization in general (see §10.3.3).

Because the TSP is a hard problem there has been a lot of work for almost every exact as well as every heuristic method or principle applied to this problem (for excellent surveys on this see, e.g., the books by Lawler et al. (1985), Reinelt (1994), Gutin and Punnen (2002)). Local search approaches are very effective especially regarding real-world and large scale instances of the TSP (see, e.g., Johnson and McGeoch (1997)).

The TSP offers lessons in the art of modeling linear problems. It can be modeled, e.g., with a number of constraints that is linear in the problem size, but a cubic number of variables or a linear number of variables and exponential growth in the number of constraints; see, e.g., Gouveia and Voß

(1995), Padberg and Sung (1991). From a teaching perspective one may learn a lot once the question has been answered, "what makes a TSP a TSP?" (see Sniedovich and Voß (2005)).

The TSP was also used as a very good starting point for various modifications and extensions such as the time constrained TSP or the time-dependent TSP. An important literature concerns the vehicle routing problem where one salesman is replaced (conceptually) by many vehicles perhaps not starting at one and the same depot but at more than one depot; see, e.g., the collection in Toth and Vigo (2002). Extensions may be considered in the same spirit as we have seen it above for the TSP (e.g., with time windows). Finally, we arrive at vehicle scheduling. Another research area is the problem of simultaneously planning production and transportation as proposed by Daskin (1985). Examples from this literature include Blumenfeld et al. (1991), van Buer et al. (1999) and van Roy (1989).

Transportation issues are often linked with locational decisions as well. Usually the production and distribution locations are assumed to have been optimized by a decision process that operates on a longer time scale than the models considered in this book. Nevertheless, as locational decisions and building a distribution network greatly influences transportation costs, these problems are closely related to simultaneous production planning and supply chain design. As an example we borrow a model from Domschke and Voß (1990).

In this model we assume an enterprise which produces P products or SKUs which are used at N different markets. External demand for SKU k at market j is assumed to be $D(j,k)$. While in the transportation model presented above we had decision variables $t_{i,j}$ indicating the quantity of a homogenous good to ship from i to j, these may be modified for handling various products $k = 1, \ldots, P$ by adding one more index: $t_{i,j,k}$. Cost values are then given by $T(i,j,k)$, correspondingly. This leads to the following set of constraints guaranteeing that all demands are fulfilled.

$$\sum_{i=1}^{L} t_{i,j,k} = D(j,k) \qquad j = 1, \ldots, N, \;\; k = 1, \ldots, P$$

In this model we further apply some nice way of modeling non-linearities by using, other than SOS2, some piecewise linear functions. Assuming production within certain boundaries we consider a linear function that takes into account (similar to our discussion of marginal transportation discounts in §6.4.4) economies of scale. More specifically let us define $\lambda(i,k)$ and $B(q,i,k)$ indicating a minimum and a maximum amount of production allowed at facility i for SKU k. The values $\lambda(i,k)$ can be thought of as strategic lower limits according to our first abstract optimization model on page 3. Within these boundaries there are piecewise linear functions indicating production costs. To illustrate the concept we first assume that we have $q = 2$ cost functions; see Figure 10.11. The change between functions happens at some amount of production, say $B(1,i,k)$. That is, any amount of SKUs produced between $\lambda(i,k)$ and $B(1,i,k)$ has a cost of $C(1,i,k)$ per unit while any additional SKU above $B(1,i,k)$ is produced at a different cost rate between $B(1,i,k)$ and $B(2,i,k)$ and accounts for a per unit cost of $C(2,i,k)$. We may view this as different cost rates of production with corresponding cost functions. Naturally, this may be generalized to the case where we have more than one intermediate boundary and a correspondingly enlarged number of piecewise linear functions for production costs.

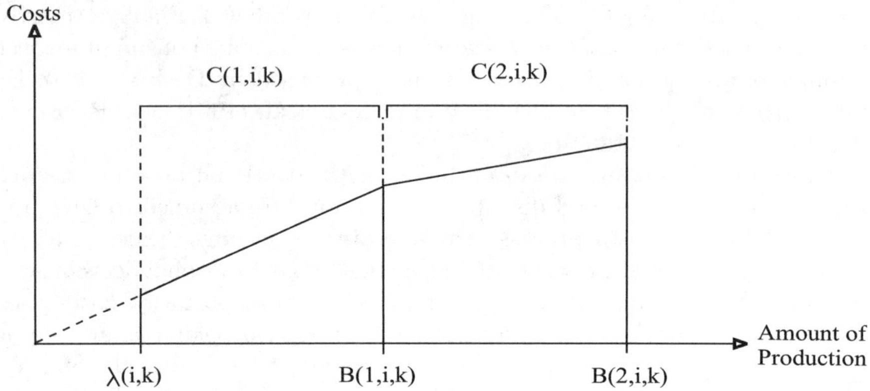

Fig. 10.11. Cost Function

The production of an SKU is automatically allocated to the cost rates. Variables $x^s_{i,k}$ indicate the number of SKUs k produced at location i at cost rate s. Hence, to compute the production quantity for an SKU, the $x^s_{i,k}$ must be summed over all rates. Binary variables $z^s_{i,k}$ indicate whether production of k at i is allowed at cost rate s. We have the following constraints that set these variables correctly depending on the overall number of SKUs produced. If production takes place, it has to be at least $\lambda(i,k)$, and with $B(0,i,k)=0$ production variables are set as follows.

$$\begin{aligned} \lambda(i,k)z^1_{i,k} &\leq x^1_{i,k} && \forall i,\ \forall k \\ (B(s-1,i,k)-B(s-2,i,k))z^s_{i,k} &\leq x^{s-1}_{i,k} && \forall i,\ \forall k,\ \ s=2,\dots,q \\ x^s_{i,k} &\leq (B(s,i,k)-B(s-1,i,k))z^s_{i,k} && \forall i,\ \forall k,\ \ s=1,\dots,q \end{aligned}$$

Production is going to take place at any of at most L locations depending on whether we use these locations or not. Whenever we open up a facility at a specific location, say i, then this implies some fixed costs of $F(i)$. Using variables y_i indicating whether to open a facility at location i or not, these location variables may also be used to allow the initialization of production indicator variables:

$$z^1_{i,k} \leq y_i \qquad i=1,\dots,m,\ \ k=1,\dots,P$$

P	Number of SKUs
L	Number of possible locations
N	Number of markets
q	Number of production cost rates
$D(j,k)$	External demand for SKU j at market k
$F(i)$	Fixed cost for opening a facility at location i
$C(s,i,k)$	Production cost rate s for SKU k at location i
$B(s,i,k)$	Cost function boundary for rate s and SKU k at location i
$B(0,i,k)$	$= 0$ (dummy cost function boundary)
$\lambda(i,k)$	Lower bound for useful production of SKU k at location i
$T(i,j,k)$	Transportation costs for SKU k between location i and market j

Table 10.1. Data for the Location-Allocation Model

y_i	Location variable
$z^s_{i,k}$	Production indicator variable for rate s and SKU k at location i
$x^s_{i,k}$	Production quantity at cost rate s for SKU k at location i
$t_{i,j,k}$	Transportation variable for SKU k between location i and market j

Table 10.2. Variables for the Location-Allocation Model

In the objective function

$$\text{minimize:} \quad \sum_{i=1}^{L} F(i)y_i + \sum_{i=1}^{L}\sum_{j=1}^{N}\sum_{k=1}^{P} T(i,j,k)t_{i,j,k} + \sum_{s=1}^{q}\sum_{i=1}^{L}\sum_{k=1}^{P} C(s,i,k)x^s_{i,k}$$

we consider the fixed costs for opening facilities, $\sum_{i=1}^{L} F(i)y_i$, the sum of all transportation costs, $\sum_{i=1}^{L}\sum_{j=1}^{N}\sum_{k=1}^{P} T(i,j,k)t_{i,j,k}$, as well as the production costs, $\sum_{s=1}^{q}\sum_{i=1}^{L}\sum_{k=1}^{P} C(s,i,k)x^s_{i,k}$. Using data and variables as shown in Tables 10.1 and 10.2 we can now summarize our model in Figure 10.12.

10.5 Optimization

A wealth of mathematics literature is devoted to one or the other aspect of optimization. The research literature typically divides optimization problems along a number of lines.

- We may distinguish between deterministic and stochastic models based on the characteristics of the data that we are provided with.
- Up to this point we have considered only models with one objective function but frequently *multi-criteria* models are considered. These are models with more than one objective function. If we want to find an optimal solution to a model with more than one objective, then we might have to provide data concerning the relative importance of the objectives.
- We may distinguish between linear and non-linear models based on the characteristics of the constraints and the objective function.

Minimize:

$$\sum_{i=1}^{L} F(i)y_i + \sum_{i=1}^{L}\sum_{j=1}^{N}\sum_{k=1}^{P} T(i,j,k)t_{i,j,k} + \sum_{s=1}^{q}\sum_{i=1}^{L}\sum_{k=1}^{P} C(s,i,k)x_{i,k}^{s}$$

subject to:

$$\begin{array}{lll}
\sum_{i=1}^{n} t_{i,j,k} & = D(j,k) & \forall j,\ \forall k \\
\sum_{j=1}^{m} t_{i,j,k} & = \sum_{s=1}^{q} x_{i,k}^{s} & \forall i,\ \forall k \\
z_{i,k}^{1} & \leq y_i & \forall i,\ \forall k \\
\lambda(i,k)z_{i,k}^{1} & \leq x_{i,k}^{1} & \forall i,\ \forall k \\
x_{i,k}^{s-1} & \geq (B(s-1,i,k) - B(s-2,i,k))z_{i,k}^{s} & \forall i,\ \forall k,\ s = 2,\ldots,q \\
x_{i,k}^{s} & \leq (B(s,i,k) - B(s-1,i,k))z_{i,k}^{s} & \forall i,\ \forall k,\ s = 1,\ldots,q \\
t_{i,j,k} & \geq 0 & \forall i,\ \forall j,\ \forall k \\
x_{i,k}^{s} & \geq 0 & \forall i,\ \forall k,\ s = 1,\ldots,q \\
z_{i,k}^{s} & \in \{0,1\} & \forall i,\ \forall k,\ s = 1,\ldots,q \\
y_i & \in \{0,1\} & \forall i
\end{array}$$

Fig. 10.12. Location-Allocation Model

- Further, we may make distinctions between models with only real variables, only binary variables, only integer variables or models that have mixtures of two or more types of variables.
- Regarding input data, offline models assume all input data of a problem instance as known in advance. On the other hand, there are many real-world (decision) problems where one can not assume that all input data is known beforehand. Online models cope with new data that become available dynamically, e.g., when the problem instance or the given constraints change.
- We may also make distinctions concerning the effort required theoretically in the worst case to find an optimal solution for a model. This, however, is beyond the scope of our book; see Garey and Johnson (1979) for a comprehensive treatment.

10.5.1 Exact Methods

The literature on exact methods is far too large for us to consider. Consequently, we restrict our attention only to those areas that were explicitly discussed in Chapter 8.

For our purposes the term solver describes readily available software for the solution of problems or models with certain properties. That is, there are solvers for linear programming problems (LP solver), those for mixed integer programs (MIP solver, which make special use of LP solver combined with branch and bound), and constraint programming solver.

Branch and bound is a very old idea and can also be applied to problems other than MIPs. A good discussion of early applications of branch and bound for MIPs is Geoffrion and Marsten (1972). Beale and Tomlin's proposal for SOS facilities reportedly appeared first in Beale and Tomlin (1970). And as in many other fields, branch and bound research is on-going (see, e.g., Liao (1994), Belvaux and Wolsey (2000)).

An important issue once a MIP has been solved is referred to as *sensitivity analysis*. A significant line of research has been devoted to providing methods that determine the effect on the optimal solution of changes to the input data. A related area of research concerns determining what changes to the data would change a problem from being infeasible to feasible. An extensive survey of this literature is provided by Greenberg (1998).

Constraint (logic) programming has origins in artificial intelligence; see Robinson (1965), Laurière (1978). Early applications were to problems in scheduling, e.g., by Fox and Smith (1984). Cooperation between CP and methods developed for MIPs is a relatively new and promising research area; see Hooker (1998), McAloon et al. (1998), Milano (2004).

10.5.2 Heuristic Search Methods

Heuristics for many optimization problems in production planning and supply chain management are based on the notion of greediness as introduced in §8.4.1. Especially in production planning many of these heuristics are called scheduling rules. More specifically, we may speak of a priority rule to represent the technique by which a number (a priority) is assigned to each job that has to be processed. Then jobs are sequenced according to these numbers, e.g., in increasing order. A simple example is the earliest due date rule where priority is given to jobs with an earlier due date over those with a later due date. Priority rules may be characterized as being static or dynamic. They are static, if they do not change the given priority once assigned. If some information is included into the rule that might change the priority in due course then it is called dynamic. An example is the nearest neighbor routine for solving the TSP. Starting with a single city, any as yet unvisited city can get a priority based on the distance to reach it. Then in every iteration the city with the best priority (in this case the smallest value, the nearest neighbor) among all cities not yet visited is chosen. From that city priorities are again given to all unvisited cities until a route through all cities has been found.

Priority rules for machine scheduling can be found, e.g., in Haupt (1989). An overall good starting point into the area of heuristic search is the book of Pearl (1984).

Much of the research on heuristic search literature focuses on meta-heuristics, which have been defined as follows: "A meta-heuristic is an iterative master process that guides and modifies the operations of subordinate heuristics to efficiently produce high-quality solutions. It may manipulate a complete (or incomplete) single solution or a collection of solutions at each iteration." (Voß et al. (1999), p. ix) These methods include simulated annealing, tabu search, genetic algorithms and many others. Recent surveys and collections can be found in Blum and Roli (2003), Glover and Kochenberger (2003), Ibaraki et al. (2005), Rego and Alidaee (2005), Ribeiro and Hansen (2002), Voß (2001).

One of the key aspects regarding metaheuristics in general is the interplay between intensification (concentrating the search into a specific subset of all possible solutions; one can think in terms of a region of the search space) and diversification (elaborating various diverse regions within the search space). That is, it has very often been appropriate, on one hand, to explore promising regions of the search space in a detailed manner (intensification) and, on the other hand, to lead the search into new and yet unexplored regions of the search space (diversification). Within intelligent search including the relationship between these two significant mechanisms the exploration of memory plays a most important role in ongoing research; see, e.g., Greistorfer and Voß (2005).

Simulated annealing traces its origins to computational simulation of the cooling of metals; see Kirkpatrick et al. (1983), Metropolis et al. (1953). Using certain cooling schedules, simulated annealing algorithms can be shown to converge to optimal solutions; see, e.g., Lundy and Mees (1986) or Hajek (1988). Cooling schedules used in practice usually differ significantly from those that are theoretically best, since the latter would result in impractically large computing times. For a more detailed discussion of simulated annealing and the effects of the cooling schedule see, e.g., Johnson et al. (1989). Based on experience using the algorithm as described in §8.3, we recommend the parameter settings INITPROB = 0.4, TEMPFACTOR = 0.8145, SIZEFACTOR = 4, and MINPERCENT = 2, which are slightly different from those recommended by Johnson et al. (1989). Cooling schedules that may behave superior to those offered by Johnson et al. are used in the Adaptive Simulated Annealing developed by Ingber (1993).

Genetic algorithms originated in work by Holland and others; see, e.g., Holland (1975), Fogel (1998). Readers interested in mathematical characterizations of early GA's should refer to Liepins and Vose (1992). Attempts to characterize the theoretical behavior continue. See, for example, the work of Salomon (1996) or Aytug and Koehler (1996).

The simple GA from Vose (1999) that we presented to ease exposition, as well as the stylized versions used for theoretical analysis, can be improved substantially for use in practice. For example, we recommend steady-state replacement without duplicates (see Syswerda (1989) or Davis (1991)) rather than generational replacement. The simple selection technique we gave is dominated by others such as a linear normal ranking scheme for all parents; see Whitley (1989). Many modern GAs perform a descent from each new population member (i.e., they combine the ideas of GA and local search). Genetic and evolutionary algorithms are large areas of ongoing research with many new, partially tested ideas; see, e.g., Smith et al. (1998), Bäck (1997), Reeves and Rowe (2003).

There exist several libraries for genetic algorithms. In principle, an advantage of using classic genetic algorithm libraries such as Genitor (2005) or GAlib (2005) is that no neighborhood must be specified. If the built-in genomes of a genetic algorithm library adequately represent one's problem, a user-specified objective function may be the only problem-specific code that must be written. Unfortunately, genetic algorithms without a local search component have not generally proven to be very effective. For a comprehensive overview of genetic algorithm libraries the reader is referred to Pain and Reeves (2002).

GAs are closely related to *evolutionary strategies*. Whereas the mutation operator in a GA serves to protect the search from premature loss of information, evolutionary strategies may incorporate some sort of local search procedure with self adapting parameters involved in the procedure. For some interesting insights on evolutionary algorithms the reader is referred to Hertz and Kobler (2000).

Tabu search was originally developed by Glover (1986) and has been extended in many directions as described in Glover and Laguna (1997). The flexibility and general applicability of TS has caused it to be used in conjunction with other heuristic search methods and much of the development work in TS is done as part of more general heuristic search efforts; see, e.g., Voß et al. (1999).

Recently, *scatter search* ideas established a link between early ideas from various sides – evolutionary strategies, TS and GAs. As an evolutionary approach, scatter search originated from strategies for creating composite decision rules and surrogate constraints. Scatter search is designed to operate on a set of points, called reference points, that constitute good solutions obtained from previous solution efforts. The approach systematically generates linear combinations of the reference points to create new points, each of which is mapped into an associated point that yields integer values for discrete variables. For a very comprehensive treatment of scatter search see Laguna and Martí (2003).

GRASP is usually composed of the following components: A greedy construction phase combined with a probabilistic component and a local search procedure. An adaptive mechanism is used to modify the greedy construction after each iteration. The basic concept goes back to ideas from Hart and Shogan (1987). Resende and Festa (2005) present a general bibliography of GRASP.

The research literature is full of comparisons of different heuristic search methods for various problems and it is difficult to declare one or the other method as clear winner. Nevertheless, based on our own research we believe that more intelligent approaches have advantages (e.g., advanced TS implementations over SA; see, among others, Voß (1996), Fink and Voß (1999a), Woodruff and Spearman (1992)).

One of the important research topics over the last couple of years is the development of *class libraries* and *frameworks*; see Voß and Woodruff (2002), Fink et al. (2003). The crucial problem of local search based meta-heuristics libraries is a generic implementation of heuristic approaches as reusable software components, which must operate on arbitrary solution spaces and neighborhood structures. The drawback is that the user must, in general, provide some kind of a problem/solution definition and a neighborhood structure, which is usually done using sophisticated computer languages such as C++.

An early C++ class library for heuristic optimization by Woodruff (1997) included both local search based methods and genetic algorithms. This library raised issues that illustrate both the promise and the drawbacks to the adaptable component approach. From a research perspective such libraries can be thought of as providing a concrete taxonomy for heuristic search. So concrete, in fact, that they can be compiled into machine code. This taxonomy sheds some light on the relationships between heuristic search methods for optimization and on ways in which they can be combined. Furthermore, the library facilitates such combinations as the classes in the library can be extended and/or combined to produce new search strategies.

From a practical and empirical perspective, these types of libraries provide a vehicle for using and testing heuristic search optimization. A user of the library must provide the definition of the problem specific abstractions and may systematically vary and exchange heuristic strategies and corresponding components.

We briefly mention one example from several heuristic optimization libraries from the research field, which differ, e.g., in the design concept, the chosen balance between "ease-of-use" and flexibility and efficiency, and the overall scope. All of these approaches are based on the concepts of object-oriented programming.

HotFrame, a Heuristic OpTimization FRAMEwork implemented in C++, provides both adaptable components that incorporate different meta-heuristics and an architectural description of the collaboration among these components and problem-specific complements. All typical application-specific concepts are treated as objects or classes: problems, solutions, neighbors, solution and move attributes. On the other side, meta-heuristics concepts such as different methods and their building-blocks such as tabu criteria and diversification strategies are also treated as objects. HotFrame uses genericity as the primary mechanism to make these objects adaptable. That is, common behavior of meta-heuristics is factored out and grouped in generic classes, applying static type variation. Meta-heuristics template classes are parameterized by corresponding aspects such as solution spaces and neighborhood structures.

All heuristics such as TS, SA and GA are implemented in a consistent way, which facilitates an easy embedding of arbitrary methods into application systems or as parts of more advanced/hybrid methods. Both new meta-heuristics and new applications can be added to the framework. For example, the *pilot method* of Duin and Voß (1999) is a technique based on lookahead that was readily implemented and added to HotFrame. Starting with a simple greedy algorithm such as a construction heuristic the pilot method builds primarily on the idea to look ahead for each possible local choice (by computing a so-called "pilot" solution), memorizing the best result, and performing the according move. The look ahead mechanism of the pilot method is related to increased neighborhood depths as it exploits the evaluation of neighbors at larger depths to guide the neighbor selection at depth one; see also Voß et al. (2005).

HotFrame includes built-in support for solution spaces representable by binary vectors or permutations, in connection with corresponding standard neighborhood structures, solution and move attributes, and recombination operators. Otherwise, the user may derive specialized classes from suitable built-in classes or implement corresponding classes from scratch according to a defined interface. For further information about HotFrame see Fink and Voß (1999b, 2002).

10.5.3 Progressive Hedging

We limit our discussion of the stochastic programming literature to those articles related to progressive hedging or multi-stage mixed integer problems. Readers interested in more general treatment should see Kall and Wallace (1994) or Birge and Louveaux (1997).

Progressive hedging is not the only method that has been proposed for multi-stage stochastic MIPs. Klein Haneveld and van der Vlerk (1999) provide descriptions of general formulations and solution methods for integer stochastic programs. Carøe and Tind (1997, 1998) describe two different methods for stochastic MIPs. Schultz et al. (1998) have developed a mathematically sophisticated method of finding provably optimal solutions to classes of stochastic MIPs. Jonsbråten et al. (1998) describe a class of stochastic MIPs where decisions affect the timing of information discovery along with a solution method.

Progressive hedging has been used in a number of applications reported in the literature. Mulvey and Vladimirou (1991, 1992) have reported success solving network problems. Helgason and Wallace (1991), Wallace and Helgason (1991) have reported success solving fishery problems and have suggested the use of tree based data structures for managing data of the PH progressive hedging algorithm.

Birge et al. (1995) report on the use of progressive hedging for power system optimization (although they use a linear, rather than a quadratic, penalty term). Carøe and Schultz (1999) propose the use of a relaxation that is similar to progressive hedging, but also uses a linear penalty. Both papers report good computational results.

The progressive hedging algorithm as described in §9.2.3 developed by Løkketangen and Woodruff (1996) is based on a more general algorithm proposed by Rockafellar and Wets (1991). For some basics regarding an interpretation as dual prices for the implementability constraints see, e.g., Wets (1989). For move evaluation functions and respective tabu search mechanisms associated with solving general stochastic MIPs see also the work described in detail by Løkketangen and Glover (1996). For an application of this algorithm to a classic single machine lot sizing problem see Haugen et al. (2001). The notion of integer convergence for progressive hedging is introduced by Løkketangen and Woodruff (1996). Related to the topics raised in this book we investigate the progressive hedging algorithm in Woodruff and Voß (2006). Based on the **SCPc** model we consider the case when an actor in the supply chain is faced with the potential for a major disruption. The progressive hedging algorithm is combined with a GRASP aiming at a realistic chance to solve models that explicitly consider the possibility of a "big bang" in the supply chain.

10.5.4 Simulation

Owing to its inherent modeling flexibility, simulation is often regarded as a proper means for supporting decision making, e.g., on supply chain design. Especially, discrete event simulation has been used to analyze and improve operations in logistic systems for more than two decades by now. Typical simulation tasks are to verify whether a system is able to produce the demanded output per time unit, to determine buffer-sizes, to identify bottlenecks, or to optimize control policies especially in cases when analytical tools are not at hand or somewhat not applicable (e.g, due to computation times). In discrete event simulation the state of a model changes at only a discrete, but possibly random, set of simulated points in time.

For a good textbook on simulation we refer to Law and Kelton (2000). Moreover, optimization in stochastic systems incorporating parametric (static) as well as control (dynamic) optimization asks for simulation and has been investigated to some extent; see, e.g., Gosavi (2003). A survey on available simulation software is conducted by Swain (2003).

As one example for supply chain simulation we mention van der Zee and van der Vorst (2005), who provide a brief literature survey with the aim of listing simulation model qualities essential for supporting successful decision making on supply chain design. Based on this the authors propose an object-oriented modeling framework that facilitates supply chain simulation.

10.6 Modeling

Modeling is a very broad and important topic. We have focused on the creation of mathematical models for optimization, but there are numerous alternative model forms, some of which we have briefly mentioned. The work of Pidd (2003) provides consideration of a wider view of modeling.

For a discussion of the art and science of MIP and LP modeling, the work of Williams (2000) is arguably the best. This book covers a large number of modeling concepts and considers the implications for solvability. The book is very comprehensive. For a more gentle introduction, operations research and management science textbooks such as the work of Hillier and Lieberman (2004) or Moore and Weatherford (2001) are useful and these books contain information about other facets of operations research modeling as well.

Muhanna (1993), Muhanna and Pick (1994) advocate object based approaches to the creation and management of mathematical programming models in a fashion similar in spirit to the structured methods proposed by Geoffrion (1992). Although not commercially available, the idea is compelling. By creating object classes to correspond to model components, models can be constructed more quickly and maintained more efficiently. Their work draws on concepts developed in the Object Oriented Modeling (OOM) literature (see, e.g., Booch et al. (1998)). A related idea is the merging of OOM techniques and modeling languages, particularly in the area of constraint logic programming and combinations with local search. For example, Michel and van Hentenryck (2001) and Laburthe and Caseau (1998) explore these notions.

In order to apply optimization methods to a new type of problem, corresponding models and algorithms have to be "coded" so that they are accessible to a computer. One way to achieve this is the use of a *modeling language*. Over the years substantial progress has been made in developing tools to simplify the design and implementation of models and algorithms. One of the research achievements is a considerable reduction in development time while preserving most of the efficiency of specialized software.

Modeling languages are being extended into new domains such as complementarity problems (see Ferris et al. (1999)) and stochastic linear programming (see Buchanan et al. (2002)). Extension of the modeling language domain to include combinatorial optimization problems is also the topic of on-going research. Such problems can often be specified more naturally as constraint programs than as integer programs, which can be exploited by modeling languages that have constraint programming capabilities (Fourer (1998), van Hentenryck (1999), van Hentenryck and Michel (2002)). These approaches have been quite successfully applied to problems with a significant number of logical constraints (for example, special scheduling and assignment problems). Sometimes modeling language support for a mixture of CP and MIP capabilities is the most effective means of addressing a particular problem (see, e.g., Jain and Grossmann (2001)).

Modeling languages provide very high-level algebraic and set notations to concisely express mathematical problems that can then be solved using state-of-the-art solvers. Because these modeling languages do not require specific programming skills they are readily used by a wide audience. In Chapter 7 we provided implementations of **mrp**, **MRPII**, and **SCPc** in some popular modeling languages, namely AMPL (*Algebraic Modeling Language for Mathematical Programming*; see, e.g., Fourer et al. (2002), Fourer (1998)), GAMS (*General Algebraic Modeling System*; see Bisschop and Meeraus (1982), Brooke et al. (1992)), MPL (*Mathematical Programming Language*), OPL (*Optimization Programming Language*), and Mosel (see, e.g., Colombani and Heipcke (2002), Guéret et al. (2002), Begain et al. (2001)). As we have shown, the modeling languages provide the power and flexibility to express well-known production planning models as optimization opportunities and support their extension to enterprise planning models.

Bibliography

Adenso-Díaz, B. and M. Laguna (1996). Modelling the load levelling problem in master production scheduling for mrp systems. *International Journal of Production Research 34*, 483–493.

Afentakis, P., B. Gavish, and U. Karmarkar (1984). Computationally efficient optimal solutions to the lot-sizing problem in multistage assembly systems. *Management Science 30*, 222–239.

Aggoun, L. and L. Benkherouf (2002). Filtering and predicting the cost of hidden perished items in an inventory model. *Journal of Applied Mathematics and Stochastic Analysis 15*(3), 235–245.

Asmundsson, J., R.L. Rardin, and R. Uzsoy (2002). Tractable nonlinear capacity models for aggregate production planning. Working paper, School of Industrial Engineering, Purdue University, West Lafayette.

Asmundsson, J., R.L. Rardin, and R. Uzsoy (2003). An experimental comparison of linear programming models for production planning utilizing fixed lead time and clearing functions. Working paper, School of Industrial Engineering, Purdue University, West Lafayette.

Atamtürk, A. and D.S. Hochbaum (2001). Capacity acquisition, subcontracting, and lot sizing. *Management Science 47*, 1081–1100.

Aytug, H. and G. J. Koehler (1996). Stopping criteria for finite length genetic algorithms. *INFORMS Journal on Computing 8*, 183–191.

Bäck, T. (Ed.) (1997). *Proceedings of the Seventh International Conference on Genetic Algorithms, Michigan State University, East Lansing, MI, July 19-23, 1997*. Morgan Kaufmann, San Francisco.

Baker, K.R. (1974). *Introduction to Sequencing and Scheduling*. Wiley, New York.

Baker, K.R. and G.D. Scudder (1990). Sequencing with earliness and tardiness penalties: A review. *Operations Research 38*, 22–36.

Balkhi, Z.T. and L. Benkherouf (2004). On an inventory model for deteriorating items with stock dependent and time–varying demand rates. *Computers & Operations Research 31*, 223–240.

Bartezzaghi, E., G. Spina, and R. Verganti (1994). Lead-time models of business processes. *International Journal of Operations and Production Management 14*(5), 5–20.

Bassok, Y. and R. Anupindi (1997). Analysis of supply contracts with total minimum commitment. *IIE Transactions 29*, 373–381.

Beale, E.M.L. and J.A. Tomlin (1970). Special facilities in a general mathematical programming system for non-convex problems using ordered sets of variables. In J. Lawrence (Ed.), *Proceedings of the 5th International Conference on Operations Research*. Tavistock, London, 447–454.

Begain, K., G. Bolch, and H. Herold (2001). *Practical Performance Modeling: Application of the MOSEL Language*. Kluwer, Boston.

Belenky, A.S. (1998). *Operations Research in Transportation Systems: Ideas and Schemes of Optimization Methods for Strategic Planning and Operations Management*. Kluwer, Boston.

Belouadah, H. and C.N. Potts (1994). Scheduling identical parallel machines to minimize total weighted completion time. *Discrete Applied Mathematics 48*, 201–218.

Belvaux, G. and L.A. Wolsey (2000). *bc – prod:* A specialized branch-and-cut system for lot-sizing problems. *Management Science 46*, 724–738.

Ben-Daya, M. and A. Raouf (1994). Inventory models involving lead time as a decision variable. *Journal of the Operational Research Society 45*, 579–582.

Benkherouf, L., A. Boumenir, and L. Aggoun (2003). A diffusion inventory model for deteriorating items. *Applied Mathematics and Computation 138*(1), 21–39.

Bes, C. and S.P. Sethi (1988). Concepts of forecast and decision horizons: Applications to dynamic stochastic optimization problems. *Mathematics of Operations Research 13*, 295–310.

Bhaskaran, S. and M.A. Turnquist (1990). Multiobjective transportation considerations in multiple facility location. *Transportation Research-A 24*, 139–148.

Billington, P.J., J.O. McClain, and L.J. Thomas (1983). Mathematical programming approaches to capacity-constraint mrp systems: Review, formulation and problem reduction. *Management Science 29*, 1126–1141.

Birge, J.R. and F. Louveaux (1997). *Introduction to Stochastic Programming*. Springer, New York.

Birge, J.R., S. Takriti, and E. Long (1995). Intelligent unified control of unit commitment and generation allocation. Technical Report 94-26; revised, University of Michigan, Ann Arbor, Department of Industrial and Operations Engineering.

Bisschop, J. and A. Meeraus (1982). On the development of a general algebraic modeling system in a strategic planning environment. *Mathematical Programming Study 20*, 1–29.

Bitran, G.R., E.A. Haas, and A.C. Hax (1981). Hierarchical production planning: A single stage system. *Operations Research 29*, 717–743.

Bitran, G.R., E.A. Haas, and A.C. Hax (1982). Hierarchical production planning: A two stage system. *Operations Research 30*, 232–251.

Bitran, G.R. and A.C. Hax (1977). On the design of hierarchical production planning systems. *Decision Sciences 8*, 28–54.

Blazewicz, J., K.H. Ecker, E. Pesch, G. Schmidt, and J. Weglarz (2001). *Scheduling Computer and Manufacturing Processes* (2 ed.). Springer, Berlin.

Blum, C. and A. Roli (2003). Metaheuristics in combinatorial optimization: Overview and conceptual comparison. *ACM Computing Surveys 35*(3), 268–308.

Blumenfeld, D., L. Burns, and C. Daganzo (1991). Synchronizing production and transportation schedules. *Transportation Research-B 25*, 23–37.

Booch, G., J. Rumbaugh, and I. Jacobson (1998). *The Unified Modeling Language User Guide.* Addison-Wesley, Reading.

Bowersox, D.J. (1974). *Logistical Management.* MacMillan Publishing, New York.

Bowersox, D.J., D.J. Closs, and M.B. Cooper (2002). *Supply Chain Logistics Management.* McGraw-Hill, New York.

Brooke, A., D. Kendrick, A. Meeraus, and R. Rosenthal (1992). *GAMS, A User's Guide and Tutorial.* Scientific Press, South San Francisco, CA.

Brucker, P. (2004). *Scheduling Algorithms* (4 ed.). Springer, Berlin.

Buchanan, C.S., K.I.M. McKinnon, and G.K. Skondras (2002). The recursive definition of stochastic linear programming problems within an algebraic modeling language. *Annals of Operations Research 104*, 15–32.

Buzacott, J.A. and J.G. Shantikumar (1993). *Stochastic Models of Manufacturing Systems.* Englewood Cliffs, New York.

Cachon, G.P. (2003). Supply chain coordination with contracts. In A.G. de Kok and S.C. Graves (Eds.), *Supply Chain Management: Design, Coordination and Operation*, Handbooks in Operations Research and Management Science. North-Holland, Amsterdam, 229–339.

Caramanis, M.C. and O.M. Anli (1999). Dynamic lead time modeling for JIT production planning. In *Proceedings of the IEEE International Conference on Robotics and Automation*, Volume 2. 1450–1455.

Carøe, C.C. and R. Schultz (1999). Dual decomposition in stochastic integer programming. *Operations Research Letters 24*, 37–45.

Carøe, C.C. and J. Tind (1997). A cutting-plane approach to mixed 0-1 stochastic integer programs. *European Journal of Operational Research 101*, 306–316.

Carøe, C.C. and J. Tind (1998). L-shaped decomposition of two-stage stochastic programs with integer recourse. *Mathematical Programming 83*, 451–464.

Chand, S., V.N. Hsu, and S. Sethi (2002). Forecast, solution, and rolling horizons in operations management problems: A classified bibliography. *Manufacturing & Service Operations Management 4*, 25–43.

Chandra, P. and M.M. Tombak (1992). Models for the evaluation of routing and machine flexibility. *European Journal of Operational Research 60*, 156–165.

Chen, F. (1998). Stationary policies in multiechelon inventory systems with deterministic demand and backlogging. *Operations Research 46*, 26–34.

Chen, F. (2003). Information sharing and supply chain coordination. In A.G. de Kok and S.C. Graves (Eds.), *Supply Chain Management: Design, Coordination and Operation*, Handbooks in Operations Research and Management Science. North-Holland, Amsterdam, 341–421.

Chen, F., Z. Drezner, J.K. Ryan, and D. Simchi-Levi (2000). Quantifying the bullwhip effect in a simple supply chain: The impact of forecasting, lead times, and information. *Management Science 46*, 436–443.

Cheng, T.C.E., J.N.D. Gupta, and G. Wang (2000). A review of flowshop scheduling research with setup times. *Production and Operations Management 9*, 262–282.

Chopra, S. and P. Meindl (2003). *Supply Chain Management* (2 ed.). Prentice-Hall, Upper Saddle River.

Christopher, M. (1986). *The Strategy of Distribution Management.* Heinemann Professional Publishing, London.

Christopher, M. (1999). *Logistics and Supply Chain Management: Strategies for Reducing Cost and Improving Service.* Pitman Publishing, London.

Clark, A.R. and V.A. Armentano (1995). A heuristic for a resource-capacitated multi-stage lot-sizing problem with lead times. *Journal of the Operational Research Society 46*, 1208–1222.

Clarke, G. and J.W. Wright (1964). Scheduling of vehicles from a central depot to a number of delivery points. *Operations Research 12*, 568–581.

Cohen, M.A. and H.L. Lee (1988). Strategic analysis of integrated production distribution systems: Models and methods. *Operations Research 36*, 216–228.

Colombani, Y. and S. Heipcke (2002). Mosel: An extensible environment for modeling and programming solutions. In N. Jussien and F. Laburthe (Eds.), *Proceedings of CP-AI-OR'02 Le Croisic March 2002.* 277–290.

Cooper, M.C., D.M. Lambert, and J.D. Pagh (1997). Supply chain management: More than a new name for logistics. *The International Journal of Logistics Management 8*(1), 1–14.

Cordeau, J.-F., P. Toth, and D. Vigo (1998). A survey of optimization models for train routing and scheduling. *Transportation Science 32*, 380–404.

Daganzo, C.F. (2003). *A Theory of Supply Chains*, Volume 526 of *Lecture Notes in Economics and Mathematical Systems.* Springer, Berlin.

Daskin, M.S. (1985). Logistics: An overview of the state of the art and perspectives on future research. *Transportation Research-A 19*, 383–398.

Davis, G.M. and S.W. Brown (1974). *Logistics Management.* Lexington Books, Toronto.

Davis, L. (1991). *Handbook of Genetic Algorithms.* Van Nostrand Reinhold, New York.

de Kok, A.G. and S.C. Graves (Eds.) (2003). *Supply Chain Management: Design, Coordination and Operation.* Handbooks in Operations Research and Management Science. North-Holland, Amsterdam.

de Kok, T.G. and J.C. Fransoo (2003). Planning supply chain operations: Definition and comparison of planning concepts. In A.G. de Kok and S.C. Graves (Eds.), *Supply Chain Management: Design, Coordination and Operation*, Handbooks in Operations Research and Management Science. North-Holland, Amsterdam, 597–675.

Dickersbach, J.T. (2004). *Supply Chain Management with APO.* Springer, Berlin.

Disney, S.M. and D.R. Towill (2003). Vendor–managed inventory and bullwhip reduction in a two–level supply chain. *International Journal of Operations & Production Management 23*(6), 625–651.

Domschke, W., A. Scholl, and S. Voß (1997). *Produktionsplanung* (2 ed.). Springer, Berlin.

Domschke, W. and S. Voß (1990). Ansätze zur strategischen Standort- und Produktionsplanung - ein Anwendungsbeispiel. In K.-P. Kistner, J.H. Ahrens, G. Feichtinger, J. Minnemann, and L. Streitferdt (Eds.), *Operations Research Proceedings 1989.* Berlin, Springer, 87–94.

Dorndorf, U., E. Pesch, and T. Phan-Huy (2000). A time-oriented branch-and-bound algorithm for resource-constrained project scheduling with generalised precedence constraints. *Management Science 46*, 1365–1384.

Drexl, A., B. Fleischmann, H.-O. Günther, H. Stadtler, and H. Tempelmeier (1994). Konzeptionelle Grundlagen kapazitätsorientierter PPS-Systeme. *Zeitschrift für betriebswirtschaftliche Forschung 46*, 1022–1045.

Drexl, A. and A. Kimms (1997). Lot sizing and scheduling – Survey and extensions. *European Journal of Operational Research 99*, 221–235.

Duin, C.W. and S. Voß (1999). The pilot method: A strategy for heuristic repetition with application to the Steiner problem in graphs. *Networks 34*, 181–191.

Dyckhoff, H., R. Lackes, and J. Reese (Eds.) (2004). *Supply Chain Management and Reverse Logistics.* Springer, Berlin.

Dye, C.-Y. and H.-J. Chang (2003). A replenishment policy for deteriorating items with linear trend demand and shortages when payment periods are offered. *Information and Management Sciences 14*(2), 31–45.

Enns, S.T. (2001). MRP performance effects due to lot size and planned lead time settings. *International Journal of Production Research 39*, 461–480.

Ertogral, K. and S.D. Wu (2000). Auction-theoretic coordination of production planning in the supply chain. *IIE Transactions 32*, 931–940.

Ettl, M., G.E. Feigin, G.Y. Lin, and D.D. Yao (2000). A supply network model with base-stock control and service requirements. *Operations Research 48*, 216–232.

Ferris, M.C., R. Fourer, and D.M. Gay (1999). Expressing complementarity problems in an algebraic modeling language and communicating them to solvers. *SIAM Journal on Computing 9*, 991–1009.

Fink, A. and S. Voß (1999a). Applications of modern heuristic search methods to pattern sequencing problems. *Computers & Operations Research 26*, 17–34.

Fink, A. and S. Voß (1999b). Generic metaheuristics application to industrial engineering problems. *Computers & Industrial Engineering 37*, 281–284.

Fink, A. and S. Voß (2002). HotFrame: A heuristic optimization framework. In S. Voß and D.L. Woodruff (Eds.), *Optimization Software Class Libraries.* Kluwer, Boston, 81–154.

Fink, A. and S. Voß (2003). Solving the continuous flow-shop scheduling problem by metaheuristics. *European Journal of Operational Research 151*, 400–414.

Fink, A., S. Voß, and D.L. Woodruff (2003). Metaheuristic class libraries. In F.W. Glover and G.A. Kochenberger (Eds.), *Handbook of Metaheuristics.* Kluwer, Boston, 515–535.

Fleischmann, B. (1998). Design of freight traffic networks. In B. Fleischmann, J.A.E.E. van Nunen, M.G. Speranza, and P. Stähly (Eds.), *Advances in Distribution Logistics*, Volume 460 of *Lecture Notes in Economics and Mathematical Systems.* Springer, Berlin, 55–81.

Fleischmann, B. and H. Meyr (2003). Planning hierarchy, modeling and advanced planning systems. In A.G. de Kok and S.C. Graves (Eds.), *Supply Chain Management: Design, Coordination and Operation*, Handbooks in Operations Research and Management Science. North-Holland, Amsterdam, 455–523.

Fleischmann, B. and H. Meyr (2004). Customer orientation in advanced planning systems. In H. Dyckhoff, R. Lackes, and J. Reese (Eds.), *Supply Chain Management and Reverse Logistics.* Springer, Berlin, 297–321.

Fleischmann, M. (2001). *Quantitative Models for Reverse Logistics.* Springer, Berlin.

Fleischmann, M., P. Beullens, J.M. Bloemhof-Ruwaard, and L.N. van Wassenhove (2001). The impact of product recovery on logistics network design. *Production and Operations Management 10*, 156–173.

Fogel, D.B. (Ed.) (1998). *Evolutionary Computation: The Fossil Record.* IEEE Press, New York.

Fourer, R. (1998). Extending a general-purpose algebraic modeling language to combinatorial optimization: A logic programming approach. In D.L. Woodruff (Ed.), *Advances in Computational and Stochastic Optimization, Logic Programming, and Heuristic Search.* Kluwer, Boston, 31–74.

Fourer, R., D.M. Gay, and B.W. Kernighan (2002). *AMPL: A Modeling Language for Mathematical Programming* (2 ed.). Duxbury Press, Belmont, CA.

Fowler, J.W., G.L. Hogg, and S.J. Mason (2002). Workload control in the semiconductor industry. *Production Planning & Control 13*, 568–578.

Fox, M.S. and S.F. Smith (1984). ISIS: A knowledge-based system for factory scheduling. *Expert Systems 1*, 25–49.

GAlib (2005). `http://lancet.mit.edu/ga/`. A C++ Library of Genetic Algorithm Components; last checked Oct 2005.

Ganeshan, R. and T.P. Harrison (1995). `http://silmaril.smeal.psu.edu/misc/supply_chain_intro.html`. An Introduction to Supply Chain Management; last checked Aug 2002.

Garey, M.R. and D.S. Johnson (1979). *Computers and Intractability: A Guide to the Theory of NP-Completeness.* Freeman, New York.

Gavirneni, S., R. Kapuscinski, and S. Tayur (1999). Value of information in capacitated supply chains. *Management Science 45*, 16–24.

Genitor (2005). `http://www.cs.colostate.edu/~genitor/`. The Genitor Group; last checked Oct 2005.

Geoffrion, A.M. (1992). The SML language for structured modeling: Levels 1 and 2. *Operations Research 40*, 38–57.

Geoffrion, A.M. and R.E. Marsten (1972). Integer programming: A framework and state-of-the-art survey. *Management Science 18*, 465–491.

Geunes, J. and P.M. Pardalos (2003). Network optimization in supply chain management and financial engineering: An annotated bibliography. *Networks 42*, 66–84.

Glover, F. (1986). Future paths for integer programming and links to artificial intelligence. *Computers & Operations Research 13*, 533–549.

Glover, F.W. and G.A. Kochenberger (Eds.) (2003). *Handbook of Metaheuristics.* Kluwer, Boston.

Glover, F. and M. Laguna (1997). *Tabu Search.* Kluwer, Boston.

Goetschalckx, M., C.J. Vidal, and K. Dogan (2002). Modeling and design of global logistics systems: A review of integrated strategic and tactical models and design algorithms. *European Journal of Operational Research 143*, 1–18.

Goldratt, E.M. and R.E. Fox (1986). *The Race.* North River Press, Croton-on-Hudson.

Gosavi, A. (2003). *Simulation–Based Optimization.* Kluwer, Boston.

Gouveia, L. and S. Voß (1995). A classification of formulations for the (time-dependent) traveling salesman problem. *European Journal of Operational Research 83*, 69–82.

Goyal, S.K. and B.C. Giri (2001). Recent trends in modeling of deteriorating inventory. *European Journal of Operational Research 134*, 1–16.

Graves, S.C. (1986). A tactical planning model for job shops. *Operations Research 34*, 522–533.

Graves, S.C., D.B. Kletter, and W.B. Hetzel (1998). A dynamic model for requirements planning with application to supply chain optimization. *Operations Research 46*, S35–S49.

Greenberg, H.J. (1998). An annotated bibliography for post-solution analysis in mixed integer programming and combinatorial optimization. In D.L. Woodruff (Ed.), *Advances in Computational and Stochastic Optimization, Logic Programming, and Heuristic Search.* Kluwer, Boston, 97–148.

Greistorfer, P. and S. Voß (2005). Controlled pool maintenance for metaheuristics. In C. Rego and B. Alidaee (Eds.), *Metaheuristic Optimization Via Memory and Evolution.* Kluwer, Boston, 387–424.

Guéret, C., C. Prins, and M. Sevaux (2002). *Applications of Optimization with Xpress-MP.* Dash Optimization, Northants. Translated and revised by S. Heipcke.

Guide Jr., V.D.R. and L.N. van Wassenhove (Eds.) (2003). *Business Aspects of Closed-Loop Supply Chains.* Carnegie Mellon University Press, Pittsburgh.

Gumaer, R. (1996). Beyond ERP and MRP II – Optimized planning and synchronized manufacturing. *IIE Solutions 28*(9), 32–35.

Gupta, J.N.D., A.M.A. Hariri, and C.N. Potts (1997). Scheduling a two-stage hybrid flow shop with parallel machines at the first stage. *Annals of Operations Research 69*, 171–191.

Gutin, G. and A.P. Punnen (Eds.) (2002). *The Traveling Salesman Problem and Its Variations.* Kluwer, Boston.

Hackman, S.T. and R.C. Leachman (1989). A general framework for modeling production. *Management Science 35*, 478–495.

Hajek, B. (1988). Cooling schedules for optimal annealing. *Mathematics of Operations Research 13*, 311–329.

Hall, R.W. (Ed.) (2003). *Handbook of Transportation Science* (2 ed.). Kluwer, Boston.

Hall, R.W. and M. Racer (1995). Transportation with common carrier and private fleets: System assignment and shipment frequency optimization. *IIE Transactions 27*, 217–225.

Hall, W.R. (1983). *Zero Inventories.* McGraw-Hill, Homewood,Il.

Handfield, R.B. and E.L. Nichols (1999). *Introduction to Supply Chain Management.* Prentice Hall, Upper Saddle River.

Hart, J.P. and A.W. Shogan (1987). Semi-greedy heuristics: An empirical study. *Operations Research Letters 6*, 107–114.

Haugen, K.V., A. Løkketangen, and D.L. Woodruff (2001). Progressive hedging as a meta-heuristic applied to stochastic lot-sizing. *European Journal of Operational Research 132*, 116–122.

Haupt, R. (1989). A survey of priority rule–based scheduling. *OR Spektrum 11*, 3–16.

Hax, A.C. and D. Candea (1984). *Production and Inventory Management.* Prentice-Hall, Englewood Cliffs, NJ.

Hax, A.C. and H.C. Meal (1975). Hierarchical integration of production planning and scheduling. In M.A. Geisler (Ed.), *TIMS Studies in Management*

Science, Volume 1: Logistics. North Holland/American Elsevier, New York, 53–69.

Helgason, T. and S.W. Wallace (1991). Approximate scenario solutions in the progressive hedging algorithm. A numerical study with an application to fisheries management. *Annals of Operations Research 31*, 425–444.

Hertz, A. and D. Kobler (2000). A framework for the description of evolutionary algorithms. *European Journal of Operational Research 126*, 1–12.

Hillier, F.S. and G.J. Lieberman (2004). *Introduction to Operations Research* (8 ed.). McGraw-Hill, New York.

Holland, J.H. (1975). *Adaptation in Natural and Artificial Systems.* University of Michigan Press, Ann Arbor.

Homburg, C. and C. Schneeweiss (2000). Negotiations within supply chains. *Computational & Mathematical Organization Theory 6*, 47–59.

Hooker, J.N. (1998). Constraint satisfaction methods for generating valid cuts. In D.L. Woodruff (Ed.), *Advances in Computational and Stochastic Optimization, Logic Programming, and Heuristic Search.* Kluwer, Boston, 1–30.

Hopp, W.J. and M.L. Spearman (2000). *Factory Physics* (2 ed.). Irwin McGraw-Hill, Boston.

Hopp, W.J., M.L. Spearman, and D.L. Woodruff (1990). Practical strategies for lead time reduction. *Manufacturing Review 3*(2), 78–84.

Hsu, V.N. (2000). Dynamic economic lot size model with perishable inventory. *Management Science 46*, 1159–1169.

Huang, G.Q., J.S.K. Lau, and K.L. Mak (2003). The impacts of sharing production information on supply chain dynamics: A review of the literature. *International Journal of Production Research 41*, 1483–1517.

Hung, Y.F. and R.C. Leachman (1996). A production planning methodology for semiconductor manufacturing based on iterative simulation and linear programming calculations. *IEEE Transactions on Semiconductor Manufacturing 9*, 257–269.

Hwang, J. and M.R. Singh (1998). Optimal production policies for multi-stage systems with setup costs and uncertain capacities. *Management Science 44*, 1279–1294.

Hwang, S. and R. Uzsoy (2004). A single-product dynamic lot sizing model with work in process and congestion. Technical report, School of Industrial Engineering, Purdue University, West Lafayette.

Ibaraki, T., K. Nonobe, and M. Yagiura (Eds.) (2005). *Metaheuristics: Progress as Real Problem Solvers.* Springer, New York.

Imaizumi, J., Y. Yamakoshi, M. Murakami, and S. Morito (1998). A decomposition approach to two-stage hybrid flow shop scheduling with diverging jobs. *Communications of the Operations Research Society of Japan 43*, 624–631.

Inderfurth, K., G. Lindner, and N.R. Rachaniotis (2005). Lot sizing in a production system with rework and product deterioration. *International Journal of Production Research 43*, 1355–1374.

Ingber, L. (1993). Simulated annealing: Practice versus theory. *Mathematical and Computer Modelling 18*, 29–57.

Jagannathan, R. and L.S. Juang (1998). Solution methods for material requirement planning with lot-size dependent lead times. *Annals of Operations Research 76*, 201–217.

Jain, V. and I.E. Grossmann (2001). Algorithms for hybrid MILP/CP models for a class of optimization problems. *INFORMS Journal on Computing 13*, 258–276.

Johnson, D.S., C.R. Aragon, L.A. McGeoch, and C. Schevon (1989). Optimization by simulated annealing: An experimental evaluation; Part I, Graph partitioning. *Operations Research 37*, 865–892.

Johnson, D.S. and L.A. McGeoch (1997). The traveling salesman problem: A case study. In E.H.L. Aarts and J.K. Lenstra (Eds.), *Local Search in Combinatorial Optimization.* Wiley, Chichester, 215–310.

Johnson, L.A. and D.S. Montgomery (1974). *Operations Research in Production Planning, Scheduling and Inventory Control.* Wiley, New York.

Johnson, M.E. and D.F. Pyke (Eds.) (2000). *Special Issue on Teaching Supply Chain Management,* Volume 9 (1) of *Production and Operations Management.*

Jonsbråten, T.W., R.J.-B. Wets, and D.L. Woodruff (1998). A class of stochastic programs with decision dependent random elements. *Annals of Operations Research 82*, 83–106.

Kall, P. and S.W. Wallace (1994). *Stochastic Programming.* Wiley, New York.

Kamien, M.I. and L. Li (1990). Subcontracting, coordination, flexibility, and production smoothing in aggregate planning. *Management Science 36*, 1352–1363.

Kanet, J.J. (1988). Mrp 96: Time to rethink manufacturing logistics. *Production and Inventory Management Journal 29*(2), 57–61.

Kanet, J.J. and M. Barut (2003). Problem-based learning for production and operations management. *Decision Sciences Journal of Innovative Education 1*, 99–118.

Karmarkar, U.S. (1987). Lot sizes, lead times and in-process inventories. *Management Science 33*, 409–418.

Karmarkar, U.S. (1989a). Batching to minimize flow times on parallel heterogeneous machines. *Management Science 35*, 607–613.

Karmarkar, U.S. (1989b). Capacity loading and release planning with work-in-process (WIP) and leadtimes. *Journal of Manufacturing and Operations Management 2*, 105–123.

Karmarkar, U.S. (1993a). Lot size, manufacturing lead times and utilization. Working paper no. QM8321, University of Rochester.

Karmarkar, U.S. (1993b). Manufacturing lead times, order release and capacity loading. In S.C. Graves, A. Rinnooy Kan, and P. Zipkin (Eds.), *Logistics of Production and Inventory*, Volume 4 of *Handbooks in Operations Research and Management Science*. North-Holland, Amsterdam, 287–329.

Karmarkar, U.S., S. Kekre, and S. Kekre (1985). Lot sizing in multi-item multi machine job shops. *IIE Transactions 17*, 290–297.

Katok, E., H.S. Lewis, and T.P. Harrison (1998). Lot sizing in general assembly systems with setup costs, setup times, and multiple constrained resources. *Management Science 44*, 859–877.

Katz, M.L. (1989). Vertical contractual relations. In R. Schmalensee and R.D. Willig (Eds.), *Handbook of Industrial Organization*. Elsevier, New York.

Kekre, S. and S. Kekre (1985). Work-in-process considerations in job shop capacity planning. Technical report, GSIA, Carnegie Mellon University.

Kimura, O. and H. Terada (1981). Design and analysis of pull system, a method of multi-stage production control. *International Journal of Production Research 19*, 241–253.

Kirkpatrick, S., C.D. Gelatt Jr., and M.P. Vecchi (1983). Optimization by simulated annealing. *Science 220*, 671–680.

Klein, R. (2000). *Scheduling of Resource-Constrained Projects.* Kluwer, Boston.

Klein Haneveld, W.K. and M.H. van der Vlerk (1999). Stochastic integer programming: General models and algorithms. *Annals of Operations Research 85*, 39–57.

Klose, A., M.G. Speranza, and L.N. van Wassenhove (Eds.) (2002). *Quantitative Approaches to Distribution Logistics and Supply Chain Management*, Volume 519 of *Lecture Notes in Economics and Mathematical Systems.* Springer, Berlin.

Knolmayer, G., P. Mertens, and A. Zeier (2002). *Supply Chain Management based on SAP Systems.* Springer, Berlin.

Koenig, D.T. (1994). *Manufacturing Engineering: Principles for Optimization* (2 ed.). Taylor & Francis, Washington.

Krajewski, L.J., B.E. King, L.P. Ritzman, and D.S. Wong (1987). Kanban, mrp, and shaping the manufacturing environment. *Management Science 33*, 39–57.

Kushner, H.J. (2001). *Heavy Traffic Analysis of Controlled Queueing and Communication Networks.* Springer, New York.

Kwon, O.K., C.D. Martland, and J.M. Sussman (1998). Routing and scheduling temporal and heterogeneous freight car traffic on rail networks. *Transportation Research-E 34*, 101–115.

Laburthe, F. and Y. Caseau (1998). SALSA: A language for search algorithms. In M. Maher and J.-F. Puget (Eds.), *Principles and Practice of Constraint Programming – CP98*, Volume 1520 of *Lecture Notes in Computer Science.* Springer, Berlin, 310–324.

Laguna, M. and R. Martí (2003). *Scatter Search.* Kluwer, Boston.

Lambrecht, M.R., P.L. Ivens, and N.J. Vandaele (1998). ACLIPS: A capacity and lead time integrated procedure for scheduling. *Management Science 44*, 1548–1561.

Lamming, R. (1996). Squaring lean supply with supply chain management. *International Journal of Operations & Production Management 16*(2), 183–196.

Laporte, G. and M. Gendreau (Eds.) (1995). *Freight Transportation*, Volume 61 of *Annals of Operations Research.*

Larson, P.D. and A. Halldorsson (2004). Logistics versus supply chain management: an international survey. *International Journal of Logistics 7*(1), 17–31.

Laurière, J.-L. (1978). A language and a program for stating and solving combinatorial problems. *Artificial Intelligence 10*, 29–127.

Lautenschläger, M. and H. Stadtler (1998). Modelling lead times depending on capacity utilization. Technical report, Technische Hochschule Darmstadt, Darmstadt, Germany.

Law, A.M. and W.D. Kelton (2000). *Simulation Modeling and Analysis* (3rd ed.). McGraw-Hill, Boston.

Lawler, E.L., J.K. Lenstra, A.H.G. Rinnooy Kan, and D.B. Shmoys (Eds.) (1985). *The Traveling Salesman Problem: A Guided Tour of Combinatorial Optimization.* Wiley, Chichester.

Leachman, R.C. (1993). Modeling techniques for automated production planning in the semiconductor industry. In T.A. Ciriani and R.C. Leachman (Eds.), *Optimization in Industry: Mathematical Programming and Modelling Techniques in Practice.* Wiley, New York, 1–30.

Lee, H.L. and C. Billington (1993). Material management in decentralized supply chains. *Operations Research 41*, 835–847.

Lee, H.L., V. Padmanabhan, and S. Whang (1997). Information distortion in a supply chain: The bullwhip effect. *Management Science 43*, 546–558.

Lee, H.L., K.C. So, and C.S. Tang (2000). The value of information sharing in a two-level supply chain. *Management Science 46*, 626–643.

Lee, T.S., E.M. Malstrom, S.B. Vardeman, and V.P. Petersen (1989). On the refinement of the variable lead time/constant demand lot-sizing model: The effect of true average inventory level on the traditional solution. *International Journal of Production Research 27*, 883–899.

Leisten, R. (1998). An LP-aggregation view on aggregation in multi-level production planning. *Annals of Operations Research 82*, 413–434.

Liao, C.J. (1994). A new node selection strategy in the branch-and-bound procedure. *Computers & Operations Research 21*, 1095–1101.

Liepins, G.E. and M.D. Vose (1992). Characterizing crossover in genetic algorithms. *Annals of Mathematics and Artificial Intelligence 5*, 27–34.

Logendran, R. and P. Ramakrishna (1997). A methodology for simultaneously dealing with machine duplication and part subcontracting in cellular manufacturing systems. *Computers & Operations Research 24*, 97–116.

Løkketangen, A. and F. Glover (1996). Probabilistic move selection in tabu search for zero-one mixed integer programming problems. In I.H. Osman and J.P. Kelly (Eds.), *Meta-Heuristics: Theory & Applications.* Kluwer, Boston, 467–487.

Løkketangen, A. and D.L. Woodruff (1996). Progressive hedging and tabu search applied to mixed integer (0,1) multistage stochastic programming. *Journal of Heuristics 2*, 111–128.

Lundy, M. and A. Mees (1986). Convergence of an annealing algorithm. *Mathematical Programming 34*, 111–124.

Luscombe, M. (1993). *MRP II: Integrating the Business.* Butterworth-Heinemann, Oxford.

Martin, R.K. (1987). Generating alternative mixed-integer programming models using variable redefinition. *Operations Research 35*, 820–831.

Martinich, J.S. (1997). *Production and Operations Management: An Applied Modern Approach.* Wiley, New York.

McAloon, K., C. Tretkoff, and G. Wetzel (1998). Disjunctive programming and cooperative solvers. In D.L. Woodruff (Ed.), *Advances in Computational and Stochastic Optimization, Logic Programming, and Heuristic Search.* Kluwer, Boston, 75–96.

Meloni, C., D. Pacciarelli, and M. Pranzo (2004). A rollout metaheuristic for job shop scheduling problems. *Annals of Operations Research 131*, 215–235.

Mendoza, G.R. (2003). *Transient Behavior of Stochastic Networks: Application to Production Planning with Load-Dependent Lead Times.* Ph. D. thesis, Georgia Institute of Technology.

Metropolis, N., M. Rosenbluth, A. Rosenbluth, A. Teller, and E. Teller (1953). Equation of state calculations by fast computing machines. *Journal of Chemical Physics 21*, 1087–1092.

Meyr, H. (2002). Simultaneous lotsizing and scheduling on parallel machines. *European Journal of Operational Research 139*, 277–292.

Michel, L. and P. van Hentenryck (2001). Modeler++: A modeling layer for constraint programming libraries. Technical report, Wye College (Imperial College), Ashford, Kent UK.

Milano, M. (Ed.) (2004). *Constraint and Integer Programming.* Kluwer, Boston.

Miltenburg, J. and W. Zhang (1991). A comparative evaluation of nine well-known algorithms for solving the cell formation problem in group technology. *Journal of Operations Management 10*(1), 44–72.

Minner, S. (2000). *Strategic Safety Stocks in Supply Chains*, Volume 490 of *Lecture Notes in Economics and Mathematical Systems.* Springer, Berlin.

Missbauer, H. (2002). Lot sizing in workload control systems. *Production Planning and Control 13*(7), 649–664.

Moinzadeh, K. and S. Nahmias (1988). A continuous review model for an inventory system with two supply modes. *Management Science 34*, 761–773.

Monma, C.L. and C.N. Potts (1993). Analysis of heuristics for preemptive parallel machine scheduling with batch setup times. *Operations Research 41*, 981–993.

Moore, J.H. and L.R. Weatherford (2001). *Decision Modeling with Microsoft Excel* (6 ed.). Prentice-Hall, Upper Saddle River.

Muhanna, W.A. (1993). An object-oriented framework for model management and DSS development. *Decision Support Systems 9*, 217–229.

Muhanna, W.A. and R.A. Pick (1994). Meta-modeling concepts and tools for model management: A systems approach. *Management Science 40*, 1093–1123.

Mulvey, J.M. and H. Vladimirou (1991). Applying the progressive hedging algorithm to stochastic generalized networks. *Annals of Operations Research 31*, 399–424.

Mulvey, J.M. and H. Vladimirou (1992). Stochastic network programming for financial planning problems. *Management Science 38*, 1642–1664.

Nahmias, S. (1982). Perishable inventory theory: A review. *Operations Research 30*, 680–708.

Nahmias, S. (2004). *Production and Operations Analysis* (5 ed.). McGraw-Hill / Irwin, Chicago.

Oliver, R.K. and M.D. Webber (1982). *Outlook.* Booz, Allen & Hamilton.

Orlicky, J. (1975). *Material Requirements Planning: The New Way of Life in Production and Inventory Management.* McGraw-Hill, New York.

Ornek, A.M. and P.I. Collier (1988). The determination of in-process inventory and manufacturing lead time in multi-stage production systems. *International Journal of Operations and Production Management 8*(1), 74–80.

Padberg, M. and T. Sung (1991). An analytical comparison of different formulations of the travelling salesman problem. *Mathematical Programming 52*, 315–357.

Pahl, J., S. Voß, and D.L. Woodruff (2005). Production planning with load dependent lead times: A survey. Technical report, University of Hamburg.

Pain, A.R. and C.R. Reeves (2002). Genetic algorithm optimization software class libraries. In S. Voß and D.L. Woodruff (Eds.), *Optimization Software Class Libraries.* Kluwer, Boston, 295–329.

Pearl, J. (1984). *Heuristics: Intelligent Search Strategies for Computer Problem Solving.* Addison-Wesley, Reading.

Pezzella, F. and E. Merelli (2000). A tabu search method guided by shifting bottleneck for the job shop scheduling problem. *European Journal of Operational Research 120*, 297–310.

Pidd, M. (2003). *Tools for Thinking: Modelling in Management Science* (2 ed.). Wiley, Chichester.

Pinedo, M.L. (2005). *Planning and Scheduling in Manufacturing and Services.* Springer, New York.

Reeves, C.R. and J.E. Rowe (2003). *Genetic Algorithms – Principles and Perspectives.* Kluwer, Boston.

Rego, C. and B. Alidaee (Eds.) (2005). *Metaheuristic Optimization via Memory and Evolution.* Kluwer, Boston.

Reinelt, G. (1994). *The Traveling Salesman: Computational Solutions for TSP Applications*, Volume 840 of *Lecture Notes in Computer Science.* Springer, Berlin.

Resende, M.G.C. and P. Festa (2005). An updated bibliography of GRASP. `http://www.research.att.com/~mgcr/doc/graspbib.pdf`, last checked Oct 2005.

Ribeiro, C.C. and P. Hansen (Eds.) (2002). *Essays and Surveys in Metaheuristics.* Kluwer, Boston.

Richey, R.G., S.E. Genchev, and P.J. Daugherty (2005). The role of resource commitment and innovation in reverse logistics performance. *International Journal of Physical Distribution & Logistics Management 35*(4), 233–257.

Robinson, J.A. (1965). A machine-oriented logic based on the resolution principle. *Journal of the ACM 12*, 23–41.

Rockafellar, R.T. and R.J.-B. Wets (1991). Scenarios and policy aggregation in optimization under uncertainty. *Mathematics of Operations Research 16*, 119–147.

Rohde, J. (2004). Hierarchical supply chain planning using artificial neural networks to anticipate base–level outcomes. *OR Spectrum 26*, 471–492.

Rom, W.O., O.I. Tukel, and J.R. Muscatello (2002). MRP in a job shop environment using a resource constrained project scheduling model. *Omega 30*, 275–286.

Roundy, R. (1986). A 98%-effective lot-sizing rule for a multi-product, multi-stage production/inventory system. *Mathematics of Operations Research 11*, 699–727.

Salomon, R. (1996). Re-evaluating genetic algorithm performance under coordinate rotation of benchmark functions – A survey of some theoretical and practical aspects of genetic algorithms. *Biosystems 39*, 263–278.

Sana, S., S.K. Goyal, and K.S. Chaudhuri (2004). A production-inventory model for a deteriorating item with trended demand and shortages. *European Journal of Operational Research 157*, 357–371.

Santoso, T., S. Ahmed, M. Goetschalckx, and A. Shapiro (2005). A stochastic programming approach for supply chain network design under uncertainty. *European Journal of Operational Research 167*, 96–115.

Savaskan, C., S. Bhattacharya, and L.N. van Wassenhove (2004). Closed-loop supply chain models with product remanufacturing. *Management Science 50*, 239–252.

Schneeweiss, C.A. (1999). *Hierarchies in Distributed Decision Making.* Springer, New York.

Schoemig, A.K. (1999). On the corrupting influence of variability in semiconductor manufacturing. In P.A. Farrington, H.B. Nembhard, D.T. Sturrock, and G.W. Evans (Eds.), *Proceedings of the 1999 Winter Simulation Conference.* 837–842.

Schonberger, R.J. (1986). *World Class Manufacturing: The Lessons of Simplicity Applied.* The Free Press, New York.

Schultz, R., L. Stougie, and M.H. van der Vlerk (1998). Solving stochastic programs with integer recourse by enumeration: A framework using Gröbner basis reductions. *Mathematical Programming 83*, 229–252.

Sethi, S.P., H. Yan, and H. Zhang (2005). *Inventory and Supply Chain Management with Forecast Updates.* Springer, New York.

Shapiro, J.F. (1999). Bottom–up vs. top–down approaches to supply chain modeling. In S. Tayur, R. Ganeshan, and M. Magazine (Eds.), *Quantitative Models for Supply Chain Management.* Kluwer, Boston, 737–759.

Shapiro, J.F. (2001a). Modeling and IT perspectives on supply chain integration. *Information Systems Frontiers 3*, 455–464.

Shapiro, J.F. (2001b). *Modeling the Supply Chain.* Duxbury, Pacific Grove.

Shingo, S. (1985). *A Revolution in Manufacturing: The SMED System.* Productivity Press, Cambridge. Translated by A.P. Dillon.

Simchi-Levi, D., P. Kaminsky, and E. Simchi-Levi (2002). *Designing and Managing the Supply Chain* (2 ed.). Irwin McGraw-Hill, Boston.

Simpson, N.C. and S.S. Erenguc (1996). Multiple-stage production planning research: History and opportunities. *International Journal of Operations and Production Management 16*(6), 25–40.

Smith, G.D, N.C. Steele, and R.F. Albrecht (Eds.) (1998). *Artificial Neural Nets and Genetic Algorithms. Proceedings of the International Conference in Norwich, UK, 1997.* Springer, New York.

Sniedovich, M. and S. Voß (2005). What makes the TSP a TSP? Technical report, University of Melbourne.

Sobel, M.J. and R.Q. Zhang (2001). Inventory policies for systems with stochastic and deterministic demand. *Operations Research 49*, 157–162.

Solomon, M.M. (1987). Algorithms for the vehicle routing and scheduling problems with time window constraints. *Operations Research 35*, 254–265.

Spearman, M.L. and W.J. Hopp (1998). Teaching operations management from a science of manufacturing. *Production and Operations Management 7*(2), 132–145.

Spearman, M.L., W.J. Hopp, and D.L. Woodruff (1989). A hierarchical control architecture for CONWIP production systems. *Journal of Manufacturing and Operations Management 2*, 147–171.

Spearman, M.L., D.L. Woodruff, and W.J. Hopp (1990). CONWIP: A pull alternative to Kanban. *International Journal of Production Research 28*, 879–894.

Srinivasan, A., M. Carey, and T.E. Morton (1988). Resource pricing and aggregate scheduling in manufacturing systems. Working paper, (revised Dec 1990), GSIA.

Stadtler, H. (2000). Improved rolling schedules for the dynamic single-level lot-sizing problem. *Management Science 46*, 318–326.

Stadtler, H. (2003). Multilevel lot sizing with setup times and multiple constrained resource: Internally rolling schedules with lot-sizing windows. *Operations Research 51*, 487–502.

Stadtler, H. (2005). Supply chain management and advanced planning – basics, overview and challenges. *European Journal of Operational Research 163*, 575–588.

Stadtler, H. and C. Kilger (Eds.) (2005). *Supply Chain Management and Advanced Planning* (3 ed.). Springer, Berlin.

Stevens, G.C. (1989). Integrating the supply chain. *International Journal of Physical Distribution & Logistics Management 19*, 3–8.

Suerie, C. and H. Stadtler (2003). The capacitated lot-sizing problem with linked lot sizes. *Management Science 49*, 1039–1054.

Swain, J.J. (2003). Simulation reloaded: Sixth biennial survey of discrete–event software tools. *OR/MS Today 30*(4), 46–57.

Syswerda, G. (1989). Uniform crossover in genetic algorithms. In J.D. Schaffer (Ed.), *Proceedings of the 3rd International Conference on Genetic Algorithms.* Morgan Kaufmann, San Mateo, 2–9.

Tardif, V. and M.L. Spearman (1997). Diagnostic scheduling in finite-capacity production environments. *Computers & Industrial Engineering 32*, 867–878.

Tayur, S., R. Ganeshan, and M. Magazine (Eds.) (1999). *Quantitative Models for Supply Chain Management.* Kluwer, Boston.

Tempelmeier, H. and M. Derstroff (1996). A Lagrangean-based heuristic for dynamic multilevel multiitem constrained lotsizing with setup times. *Management Science 42*, 738–757.

Thonemann, U.W. (2002). Improving supply-chain performance by sharing advance demand information. *European Journal of Operational Research 142*, 81–107.

Tirole, J. (1988). *The Theory of Industrial Organization.* MIT Press, Cambridge.

Toth, P. and D. Vigo (Eds.) (2002). *The Vehicle Routing Problem.* Society for Industrial & Applied Mathematics, SIAM, Philadelphia.

Trigeiro, W.W., L.J. Thomas, and J.O. McClain (1989). Capacitated lot sizing with setup times. *Management Science 35*, 353–366.

Tsay, A.A., S. Nahmias, and N. Agrawal (1999). Modeling supply chain contracts: A review. In S. Tayur, R. Ganeshan, and M. Magazine (Eds.), *Quantitative Models for Supply Chain Management.* Kluwer, Boston, 299–336.

Uzsoy, R., C. Lee, and L. Martin-Vega (1994). A review of production planning and scheduling models in the semiconductor industry part ii: shop-floor control. *IIE Transactions 26*(5), 44–55.

Vaessens, R.J.M., E.H.L. Aarts, and J.K. Lenstra (1996). Job shop scheduling by local search. *INFORMS Journal on Computing 8*, 302–317.

van Buer, M.G., D.L. Woodruff, and R.T. Olson (1999). Solving the medium newspaper production/distribution problem. *European Journal of Operational Research 115*, 237–253.

van der Zee, D.J. and J.G.A.J. van der Vorst (2005). A modeling framework for supply chain simulation: Opportunities for improved decision making. *Decision Sciences 36*, 65–95.

van Hentenryck, P. (1999). *The OPL Optimization Programming Language.* MIT Press, Cambridge. With contributions by I. Lustig, L. Michel, and J.-P. Puget.

van Hentenryck, P. and L. Michel (2002). The modeling language OPL – A short overview. In S. Voß and D.L. Woodruff (Eds.), *Optimization Software Class Libraries.* Kluwer, Boston, 263–294.

van Mieghem, J.A. (1999). Coordinating investment, production, and subcontracting. *Management Science 45*, 954–971.

van Roy, T.J. (1989). Multi-level production and distribution planning with transportation fleet optimization. *Management Science 35*, 1443–1453.

Vendemia, W.G., B.E. Patuwo, and M.S. Hung (1995). Evaluation of lead time in production/inventory systems with non-stationary stochastic demand. *Journal of the Operational Research Society 46*, 221–233.

Venugopal, V. (1999). Soft-computing-based approaches to the group technology problem: A state-of-the-art review. *International Journal of Production Research 37*, 3335–3357.

Vollmann, T.E., W.L. Berry, and D.C. Whybark (1988). *Manufacturing Planning and Control Systems* (2 ed.). Dow Jones-Irwin, Homewood.

Vose, M.D. (1999). *The Simple Genetic Algorithm: Foundations and Theory.* MIT Press, Boston.

Voß, S. (1996). Dynamic tabu search strategies for the traveling purchaser problem. *Annals of Operations Research 63*, 253–275.

Voß, S. (2001). Meta-heuristics: The state of the art. In A. Nareyek (Ed.), *Local Search for Planning and Scheduling*, Volume 2148 of *Lecture Notes in Artificial Intelligence.* Springer, Berlin, 1–23.

Voß, S., A. Fink, and C. Duin (2005). Looking ahead with the pilot method. *Annals of Operations Research 136*, 285–302.

Voß, S., S. Martello, I.H. Osman, and C. Roucairol (Eds.) (1999). *Meta-Heuristics: Advances and Trends in Local Search Paradigms for Optimization.* Kluwer, Boston.

Voß, S. and G. Schneidereit (2002). Interdependencies between supply contracts and transaction costs. In S. Seuring and M. Goldbach (Eds.), *Cost Management in Supply Chains.* Physica, Heidelberg, 253–272.

Voß, S. and D.L. Woodruff (2000). Supply chain planning: Is mrp a good starting point? In H. Wildemann (Ed.), *Supply Chain Management.* TCW, München, 177–203.

Voß, S. and D.L. Woodruff (Eds.) (2002). *Optimization Software Class Libraries.* Kluwer, Boston.

Voß, S. and D.L. Woodruff (2004). A model for multi-stage production planning with load dependent lead times. In R.H. Sprague (Ed.), *Proceedings of the 37th Annual Hawaii International Conference on System Science.* IEEE, Piscataway, DTVEA03 1–9.

Wagner, H.M. and T.M. Whitin (1958). Dynamic version of the economic lot size model. *Management Science 5*, 89–96.

Wallace, S.W. and T. Helgason (1991). Structural properties of the progressive hedging algorithm. *Annals of Operations Research 31*, 445–456.

Wallace, T.F. (1990). *MRP II: Making it Happen.* Oliver Wight Limited Publications, Essex Junction, VT.

Wang, C., L. Xu, X. Liu, and X. Qin (2005). ERP research, development and implementation in China: An overview. *International Journal of Production Research 43*, 3915–3932.

Wets, R.J.-B. (1989). The aggregation principle in scenario analysis and stochastic optimization. In S.W. Wallace (Ed.), *Algorithms and Model Formulations in Mathematical Programming.* Springer, Berlin, 91–113.

Whitley, D. (1989). The GENITOR algorithm and selection pressure: Why rank-based allocation of reproductive trials is best. In J.D. Schaffer (Ed.), *Proceedings of the 3rd International Conference on Genetic Algorithms.* Morgan Kaufmann, San Mateo, 116–121.

Whittemore, A.S. and S.C. Saunders (1977). Optimal inventory under stochastic demand with two supply options. *SIAM Journal on Applied Mathematics 32*, 293–305.

Wight, O.W. (1974). *Production and Inventory Management in the Computer Age.* Cahners, Boston, MA.

Williams, H.P. (2000). *Model Building in Mathematical Programming* (4 ed.). Wiley, Chichester.

Wolsey, L.A. (1997). MIP modelling of changeovers in production planning and scheduling problems. *European Journal of Operational Research 99*, 154–165.

Woodruff, D.L. (1997). A class library for heuristic search optimization. *INFORMS Computer Science Technical Section Newsletter 18*(2), 1–5.

Woodruff, D.L. and M.L. Spearman (1992). Sequencing and batching for two classes of jobs with deadlines and setup times. *Production and Operations Management 1*(1), 87–102.

Woodruff, D.L. and S. Voß (2006). Planning for a big bang in a supply chain: Fast hedging for production indicators. In R.H. Sprague (Ed.), *Proceedings of the 39th Annual Hawaii International Conference on System Science.* IEEE, Piscataway, DTIDSE03 1–6.

Wu, S.D. and H. Golbasi (2004). Multi-item, multi-facility supply chain planning: models, complexities, and algorithms. *Computational Optimization and Applications 28*, 325–356.

Zhao, X., F. Lai, and S. Young (2002). A study of manufacturing resources planning (MRPII) implementation in China. *International Journal of Production Research 40*, 3461–3478.

Zhao, X., J. Xie, and J. Leung (2002). The impact of forecasting model selection on the value of information sharing in a supply chain. *European Journal of Operational Research 142*, 321–344.

Zijm, W.H.M. and R. Buitenhek (1996). Capacity planning and leadtime management. *International Journal of Production Economics 46/47*, 165–179.

Zipkin, P.H. (1986). Models for design and control of stochastic, multi-item batch production systems. *Operations Research 34*, 91–104.

Index

Printing and Binding: Strauss GmbH, Mörlenbach